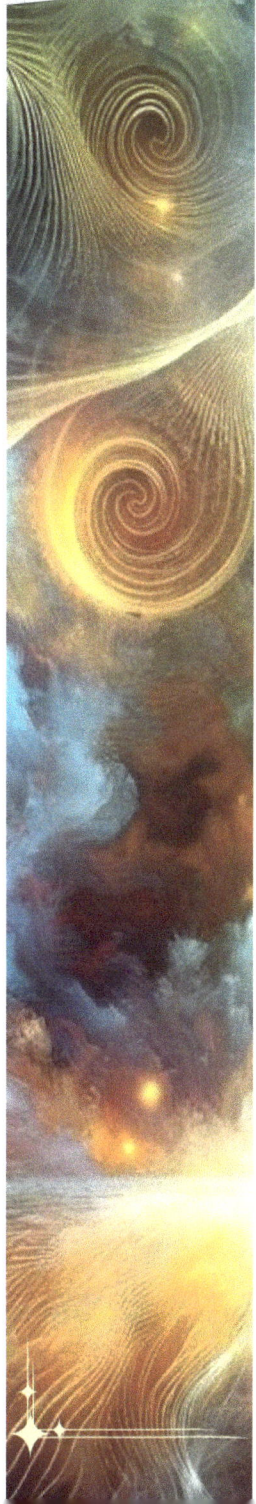

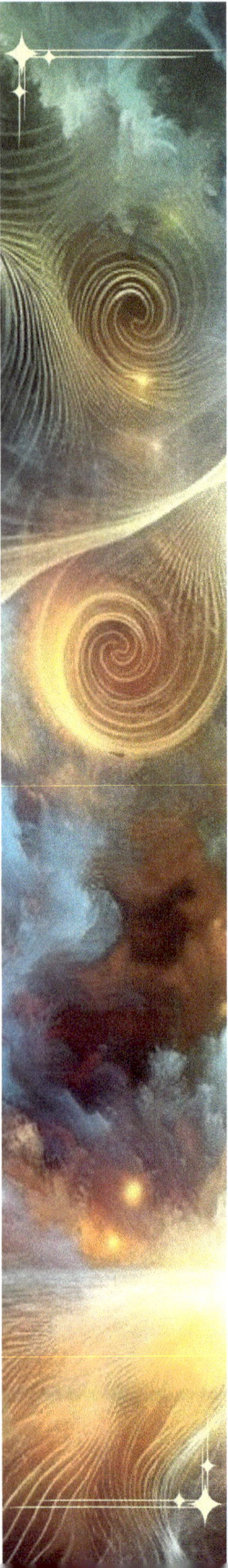

7ESLA AND THE FUTURE OF ENERGY MEDICINE

Dr. Constance Santego
Maximillian Enterprises
Kelowna, BC

Tesla and the Future of Energy Medicine
Copyright © 2024 by Dr. Constance Santego.

Copy Editor & Interior Design: Dr. Constance Santego
Book Layout: ©2017 BookDesignTemplates.com
Ordering Information:
Quantity sales. Special discounts are available on quantity purchases by corporations, associations, and others. For details, contact the "Special Sales Department" at the address above.

Trade Paperback ISBN: 978-1-990062-16-2
Ebook ISBN 978-1-990062-17-9
Created and published In Canada. Printed and bound in the United States of America

First Edition
Published by Maximillian Enterprises
Kelowna, BC
Canada
www.constancesantego.ca

*Dedicated to all the
Pioneers of Energy Medicine.*

"IF YOU WANT TO FIND THE SECRETS OF THE UNIVERSE, THINK IN TERMS OF ENERGY, FREQUENCY, AND VIBRATION."
—NIKOLA TESLA

ALSO BY DR. CONSTANCE SANTEGO

FICTION

The Nine Spiritual Gifts Series:
Journey of a Soul – (Vol. 1 Michael)
Language of a Soul – (Vol. 2 Gabriel)
Prophecy of a Soul – (Vol. 3 Bath Kol)
Healing of a Soul – (Vol. 4 Raphael)
Miracles of a Soul – (Vol. 5 Hamied)

NON-FICTION

The Intuitive Life, The Gift of Prophecy,
Third Edition
Fairy Tales, Dreams and Reality… Where Are
You On Your Path? Second Edition
Your Persona… The Mask You Wear
Angelic Lifestyle, A Vibrant Lifestyle
Angelic Lifestyle 42-Day Energy Cleanse
Archangel Michael's Soul Retrieval Guide

SECRETS OF A HEALER, SERIES:

Magic of Aromatherapy (Vol. I)
Magic of Reflexology (Vol. II)
Magic of The Gifts (Vol. III)
Magic of Muscle Testing (Vol. IV)
Magic of Iridology (Vol. V)
Magic of Massage (Vol. VI)
Magic of Hypnotherapy (Vol. VII)
Magic of Reiki (Vol. VIII)
Magic of Advanced Aromatherapy (Vol. IX)
Magic of Esthetics (Vol. X)

FOR CHILDREN

I am big tonight. I don't need the light!

Contents

"THE DOCTOR OF THE FUTURE WILL GIVE NO MEDICATION, BUT WILL INTEREST HIS PATIENTS IN THE CARE OF THE HUMAN FRAME, IN DIET AND IN THE CAUSE AND PREVENTION OF DISEASE."
—THOMAS EDISON

Foreword

In the realm of innovation and discovery, few names shine as brightly as that of Nikola Tesla. A visionary far ahead of his time, Tesla's contributions to the world of electricity and electromagnetism have shaped the very fabric of modern life. Yet, beyond the alternating current and the wireless transmissions lies Tesla's lesser-known vision—a vision of a world empowered by technology and an understanding of the subtle, invisible forces that govern health and healing. This book is an exploration of that vision, a bridge between Tesla's pioneering work and the burgeoning field of energy medicine.

As we stand on the cusp of a new healthcare era that acknowledges the interplay between energy and well-being, it becomes imperative to revisit Tesla's insights and dreams. The foreword, penned by a leading expert in the field of energy medicine, sets the stage for a journey into the heart of healing, inviting readers to open their minds to the possibilities that lie at the intersection of science, health, and the unseen energies that surround us.

Preface

At the heart of this book lies a convergence of genius and vision, a narrative where the groundbreaking work of Nikola Tesla meets the transformative potential of energy medicine. This is not merely a story of science and technology, nor is it solely a tale of healing and wellness. It is, instead, a voyage across the boundaries of what we know into the realms of what we can imagine and achieve.

Nikola Tesla, a figure synonymous with invention and electricity, spent his life pushing the limits of understanding and capability. His dreams of wireless energy transmission, free and accessible for all, were far ahead of his time. Today, we stand on the brink of realizing a different aspect of Tesla's vision: the application of energy not just to power our machines but to heal our bodies and minds.

This book emerges from a profound respect for Tesla's work and a deep belief in the untapped potential of energy medicine. As a bridge between Tesla's innovations and the holistic healing arts, it is written for those who dare to dream of a future where healing is holistic, integrative, and deeply attuned to the energies that compose and surround us.

The inspiration for "Tesla and the Future of Energy Medicine" came from a realization that, while we have advanced technologically in many ways since Tesla's time, our approach to health and healing still has much to evolve. By exploring the intersection of Tesla's electromagnetic theories and the principles of energy medicine, this book aims to open minds to the possibility of a new paradigm in healthcare—one that is as concerned with the energetic as it is with the physical.

Throughout these pages, you will find an exploration of Tesla's life and work, an overview of the principles of energy medicine, and a visionary look at how these fields might intersect to inform new healing modalities. It is a journey that traverses the past, examines the present, and dreams of the future.

As you read, I invite you to open your heart and mind to the possibilities that lie within these concepts. Whether you are a practitioner of energy medicine, a student of science, or simply someone curious about the potential for a healthier, more harmonious world, there is something in this book for you.

The journey ahead is as exciting as it is uncertain, but it is a path worth exploring. Together, let us step into the future of healing, guided by the legacy of Nikola Tesla and propelled by our collective desire for a world where energy medicine plays a pivotal role in our wellbeing.

Welcome to "Tesla and the Future of Energy Medicine."

With warmth and anticipation,

Dr. Constance Santego

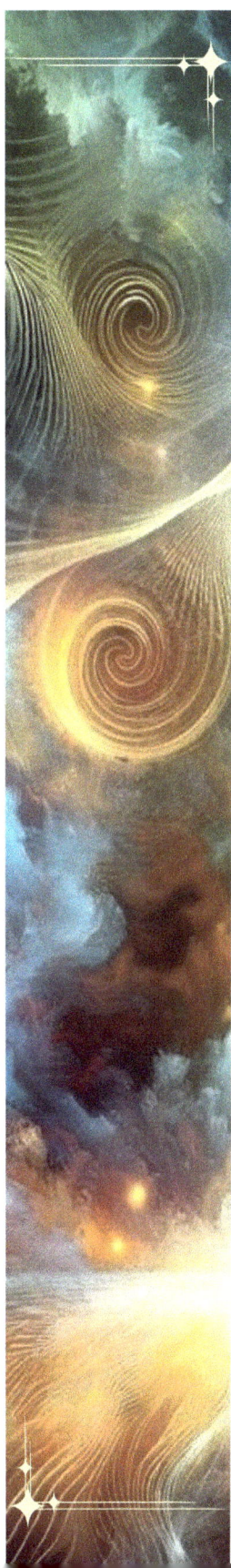

Note to Reader

*D*ear Reader,

Welcome to "Tesla and the Future of Energy Medicine," a journey that bridges the pioneering legacy of Nikola Tesla with the evolving field of energy medicine. This book is designed to illuminate the connections between Tesla's groundbreaking work and the principles that underpin modern holistic healing practices. As you embark on this exploration, here are a few notes to enhance your reading experience:

1. Open Mind: The concepts and ideas presented in this book traverse the realms of science, spirituality, and innovation. I encourage you to approach them with an open mind and curiosity, ready to explore the possibilities that lie at the intersection of Tesla's visions and contemporary health practices.
2. Interdisciplinary Approach: This book integrates insights from physics, biology, history, and alternative medicine. A basic understanding of these areas may enrich your comprehension, but it's structured to be accessible even to those with no prior expertise.
3. Practical Insights: While delving into theoretical concepts, we also focus on

practical applications and implications for health and wellness. Look for highlighted sections that offer ways to apply the principles discussed in your personal or professional life.

4. Engagement and Reflection: Each chapter is designed not only to inform but also to engage. Reflection questions and thought experiments are included to encourage you to ponder how Tesla's insights and energy medicine can impact future healthcare paradigms.

5. Further Exploration: At the end of this book, resources for further exploration are provided. Whether your interest is deepened in Tesla's work, energy medicine, or both, these resources can guide your continued journey.

6. Community and Discussion: I invite you to join the broader conversation about Tesla, energy medicine, and the future of healing. Engage with online forums, community groups, or social media platforms to share insights, ask questions, and connect with like-minded individuals.

This book represents a confluence of history, science, and the potential for a transformed approach to health and healing. It's an invitation to imagine a future where energy medicine, informed by Tesla's innovations, plays a pivotal role in achieving wellness and balance.

Thank you for choosing to embark on this journey. May you find inspiration, insight, and perhaps a glimpse into the future of healing and human potential.

Warmest regards,

Dr. Constance Santego
Author

Learning Outcome

Upon completing "Tesla and the Future of Energy Medicine," readers will:

1. Understand the Historical and Scientific Legacy of Nikola Tesla:
    - Gain comprehensive insights into Tesla's pioneering work in electricity and electromagnetism and its foundational role in modern technology and energy medicine.
2. Grasp the Principles of Energy Medicine:
    - Acquire a foundational understanding of energy medicine, including biofields, quantum healing, and the impact of electromagnetic fields on biological systems.
3. Appreciate the Interplay Between Science and Holistic Healing:
    - Recognize the intersections between Tesla's scientific principles and holistic approaches to health, highlighting the potential for integrative healing practices.
4. Explore Innovations Inspired by Tesla:
    - Discover contemporary technologies and therapies in energy medicine that draw inspiration from Tesla's work, showcasing advancements in non-invasive healing modalities.
5. Apply Concepts of Energy Medicine to Personal Wellness:

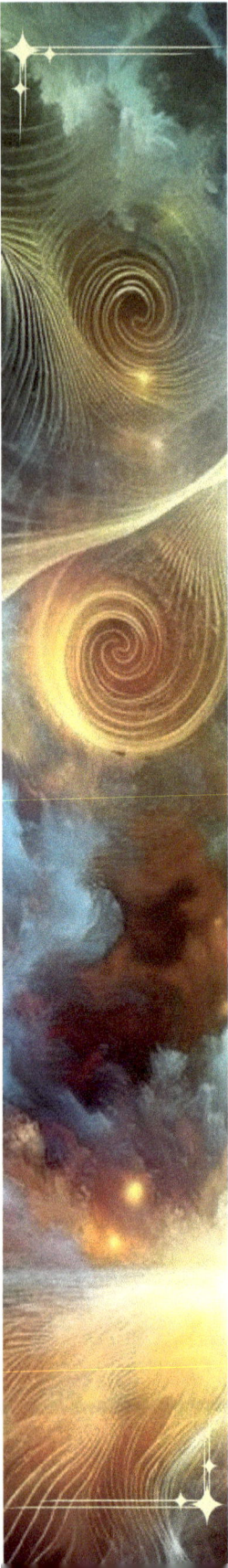

- Learn practical applications of energy medicine for enhancing personal health and wellness, including basic techniques for harnessing the body's energy systems for healing.

6. Critically Evaluate the Ethical and Environmental Implications:
   - Develop a critical perspective on the ethical considerations and environmental impacts of emerging technologies in energy medicine, fostering responsible and sustainable use.

7. Foster a Vision for the Future of Healing:
   - Cultivate a visionary outlook on the future of healthcare, imagining a world where energy medicine and Tesla's legacy contribute to a holistic and sustainable approach to healing and well-being.

8. Encourage Lifelong Learning and Curiosity:
   - Inspire continued exploration and learning about the potential of energy medicine, Tesla's contributions to science, and the evolving landscape of holistic health practices.

This book aims to equip readers with the knowledge and inspiration to explore the possibilities of energy medicine, informed by the legacy of Nikola Tesla. By bridging historical innovations with future potentials, readers will be poised to engage with and contribute to the ongoing dialogue around health, technology, and the human potential for healing.

Chapter 1

The Legacy of Nikola Tesla

THE INVENTOR OF THE MODERN WORLD

Nikola Tesla's legacy as an inventor and scientist is unparalleled, with contributions that have fundamentally shaped the modern world. Born on July 10, 1856, in Smiljan, in what is now Croatia, Tesla's genius became evident early in his life. His work spans the development of alternating current (AC) electrical systems, which enabled the widespread distribution of electricity, to groundbreaking explorations in wireless communication, radio, and even early concepts of radar technology.

Tesla's visionary projects, such as the Tesla coil, laid the groundwork for future wireless technologies. His passion for discovery led him to file over 300 patents, covering inventions that explored the realms of electricity, magnetism, and electromechanics. Tesla's AC system triumphed over Thomas Edison's direct current (DC), revolutionizing how electricity was generated, transmitted, and used, powering the world into a new era of technological advancement.

Despite facing financial difficulties and periods of obscurity, Tesla's dedication to his work never waned. His legacy is celebrated in the countless technologies that underpin modern life, from the electricity that lights our homes to the wireless devices that connect us. Tesla's contributions go beyond tangible inventions; he embodied the spirit of innovation, inspiring future generations to explore the unknown.

BEYOND ELECTRICITY: TESLA'S VISION FOR ENERGY AND HEALTH

While Tesla is most renowned for his contributions to electrical engineering, his vision extended into the potential of energy for promoting health and wellness. Tesla believed in the therapeutic potential of electromagnetic fields, a concept that presaged modern interests in energy medicine. His experiments with high-frequency currents and electromagnetic waves suggested possibilities for using energy to heal and revitalize the human body.

Tesla's insights into the electromagnetic nature of the human body anticipated the principles of bioelectromagnetics, a field that investigates the interaction between electromagnetic fields and biological entities. He hypothesized that external electromagnetic fields could have profound effects on physiological processes, potentially offering pathways to healing without invasive interventions.

One of Tesla's lesser-known inventions, the Tesla Oscillator, was a mechanical oscillator that Tesla believed could have therapeutic benefits. Although the medical community of Tesla's time did not widely embrace his ideas on energy and health, his work has inspired contemporary researchers and practitioners exploring the intersection of electromagnetic fields and healing.

Tesla's vision for energy and health was part of his broader understanding of the universe as a dynamic system of energy flows. He saw the potential for harnessing these energies not only for industrial and technological purposes but also for the fundamental enhancement of human life and well-being. This chapter explores how Tesla's pioneering work and visionary ideas lay the foundation for contemporary explorations into energy medicine, pointing towards a future where healing is informed by an understanding of the body's energetic dimensions.

Through "The Legacy of Nikola Tesla," we embark on a journey to understand how Tesla's groundbreaking innovations and visionary insights continue to influence fields beyond those he directly explored—inviting us to imagine a world where energy is not only the source of technological power but the key to health and healing.

Historical Perspectives on Energy Medicine: Tracing the Roots

Within "Tesla and the Future of Energy Medicine," Chapter 1 not only honors Nikola Tesla's legacy but also sets the stage by delving into the rich historical tapestry of energy medicine. This added section, "Historical Perspectives on Energy Medicine: Tracing the Roots," explores the profound understanding and application of energy in healing practices across ancient civilizations, providing a crucial context for Tesla's later contributions and the evolution of energy medicine into the modern era.

ANCIENT EGYPTIAN PRACTICES:

The ancient Egyptians were pioneers in recognizing the interplay between energy and health. They utilized amulets, chants, and complex rituals designed to channel cosmic and Earthly energies to promote healing and protection. Their medical papyri, such as the Ebers Papyrus, hint at an intricate knowledge of energetic anatomy, suggesting that they might have understood the concept of energy channels and fields millennia before the modern era.

GREEK PHILOSOPHICAL INSIGHTS:

The Greeks, led by figures like Pythagoras and later Hippocrates, viewed health as a state of harmony between the body, mind, and the universe. Pythagoras's theory of the harmony of the spheres proposed that celestial bodies emit their own unique frequencies, influencing human health and consciousness. Hippocrates, the father of medicine, believed in the healing power of nature ('Vis Medicatrix Naturae') and suggested that disease resulted from an imbalance of vital energies.

INDIGENOUS WISDOM:

Indigenous cultures around the world have long recognized the significance of energy in healing. Native American, Aboriginal Australian, Andean, and Shamanic traditions, to name a few, all share a deep understanding of the Earth's energies and their healing properties. These cultures employ a variety of practices, from the use of medicinal herbs and spirit guides to energy-based healing

techniques, all rooted in the belief of a vital force that connects all living things.

THE VEDIC TRADITION OF INDIA:

Rooted in texts that date back over 3,000 years, the Vedic tradition of India introduces the concept of prana (life force or vital energy) as central to health and well-being. Ayurveda, the science of life, emanates from these texts, positing that balance among the body's energies (doshas) is essential for optimal health. Techniques such as pranayama (breath control) and yoga were developed to manipulate and harmonize prana, demonstrating an intricate understanding of energy's role in physical and spiritual wellness.

TRADITIONAL CHINESE MEDICINE (TCM):

With origins tracing back to the Shang dynasty (circa 1600 BCE – 1046 BCE), Traditional Chinese Medicine embodies a comprehensive framework for understanding and influencing the body's energy flow. Central to TCM is the concept of Qi (chi), the vital life force that flows through meridians, invisible channels in the body. Practices such as acupuncture, qigong, and tai chi are designed to balance Qi, aligning the individual with the natural order of the universe to prevent and treat disease.

THE SUFI HEALING TRADITION:

The Sufi tradition, with roots in Islamic mysticism, offers a unique perspective on healing that emphasizes the heart's role as a spiritual and energetic center. Sufi healing practices involve meditation, chanting (dhikr), and breathwork to cultivate divine love and light, believed to purify the heart and soul, thereby influencing physical health. This tradition teaches that accessing and channeling spiritual energy can lead to profound healing transformations.

THE JAPANESE TRADITION OF REIKI:

Developed in the early 20th century by Mikao Usui, Reiki is based on the belief that a universal life force energy can be channeled to support healing and balance within the body. Though more modern than the other traditions mentioned, Reiki draws on ancient concepts of vital energy that are deeply embedded in Japanese culture. Practitioners use a technique called palm healing or hands-on healing, through which a "universal energy" is transferred through the palms of the practitioner to the patient to encourage emotional or physical healing.

TIBETAN MEDICINE:

Tibetan medicine is a centuries-old traditional medical system that integrates Buddhism's spiritual teachings with the medical knowledge of ancient India, Persia, and China. It emphasizes the importance of the mind-body connection and uses the body's energy system, including channels and chakras, for diagnosis and treatment. Practices such as meditation, yoga, and the use of herbal remedies aim to balance the body's energies, promoting physical, mental, and spiritual well-being.

THE POLYNESIAN CONCEPT OF MANA:

In Polynesian culture, particularly among the Maori of New Zealand, 'Mana' is a foundational concept that represents spiritual energy or healing power present in all things and people. Healing practices in these cultures revolve around the understanding and manipulation of Mana to restore balance and harmony within the individual and the community. Healers, known as tohunga, play a crucial role in diagnosing and treating illnesses by working with these energies, often through prayer, massage, and herbal medicine.

AFRICAN HEALING TRADITIONS:

Across the African continent, diverse cultures share a deep-rooted belief in the power of spiritual energy in healing. These traditions often emphasize the interconnectedness of the physical, spiritual, and natural worlds, with healers (sometimes referred to as shamans or medicine men/women) playing a crucial role in mediating these connections. Healing practices may include the use of herbs, animal products, spiritual rituals, and energy manipulation to restore balance to the individual's life force. A common theme is the use of dance and drumming to facilitate healing trances, during which healers access spiritual realms to guide energy healing processes.

THE AYAHUASCA TRADITIONS OF THE AMAZON:

Native to the Amazon rainforest, Ayahuasca is a sacred brew made from the Banisteriopsis caapi vine and other ingredients. Used for centuries by indigenous tribes for spiritual and physical healing, Ayahuasca is believed to cleanse the body and mind, allowing individuals to connect with the spiritual world and understand the energetic roots of their ailments. The ceremonies, led by a shaman, are profound spiritual experiences believed to facilitate deep healing by enabling direct encounters with the spiritual dimension, where insights into personal and collective healing can be revealed. This tradition highlights the role of plant medicine in facilitating energetic and psychological healing, showcasing another dimension of humanity's exploration of energy medicine.

This historical exploration reveals a universal thread woven through the fabric of human understanding: the recognition of energy as a fundamental component of health and healing. These ancient practices, though varied in method and philosophy, all reflect a profound insight into the energetic dimensions of life, prefiguring modern interests in biofields, quantum healing, and the therapeutic applications of electromagnetic fields.

By tracing the roots of energy medicine back through history, we gain not only a deeper appreciation for the diversity and richness of human healing traditions but also a context for understanding how Tesla's visionary work fits into a long lineage of exploration and discovery. This historical perspective enriches our journey through the book, reminding us that the quest to harness the healing power of energy is as old as humanity itself, and as new as the latest scientific research inspired by Tesla's legacy.

"EVERYTHING IN LIFE IS VIBRATION." —ALBERT EINSTEIN

Chapter 2

Foundations of Energy Medicine

THE SCIENCE BEHIND VIBRATIONS AND HEALTH

The concept of energy medicine is rooted in the understanding that every living organism emits a complex field of vibrations and electromagnetic frequencies. This chapter delves into the scientific underpinnings of how these vibrations and frequencies influence health and disease, bridging ancient wisdom with contemporary scientific research. At the heart of this exploration is the principle that the human body is not just a physical entity but a dynamic system of energy.

Recent advancements in quantum physics and biophysics have begun to provide a framework for understanding the mechanisms by which vibrations and electromagnetic fields can affect cellular function, DNA expression, and the body's inherent healing processes. Research into the effects of low-level electromagnetic fields on human health offers insights into how external energies can interact with the body's own energy fields, potentially influencing health outcomes.

This section also examines the role of resonance and entrainment in health and disease. Resonance, the phenomenon where two vibrating systems in proximity can influence each other, and entrainment, where a stronger frequency can cause another to synchronize with it, are fundamental to understanding how interventions in

energy medicine might work to restore balance and promote healing.

BIOFIELDS: THE HUMAN ENERGY ANATOMY

Building on the foundational science of vibrations and health, this chapter introduces the concept of biofields — the energy fields that both surround and penetrate the human body, often referred to as the human energy anatomy. Biofields encompass a variety of energy expressions, including but not limited to the electromagnetic fields generated by the heart and brain, as well as subtler energy systems described in traditional healing practices.

Drawing from a multidisciplinary perspective, this section explores the various models of biofields, from the detailed maps of energy channels (meridians) and centers (chakras) found in Ayurvedic and Traditional Chinese Medicine to contemporary interpretations and measurements of biofields in bioelectromagnetic research. The exploration includes discussions on how biofields are believed to regulate everything from physiological processes to emotional and psychological states, highlighting both historical perspectives and modern scientific investigations into their roles in health and healing.

Central to the discussion on biofields is the emerging evidence from research studies that support the existence of these fields and their potential impacts on health. Innovations in imaging and sensor technologies have enabled scientists to measure biofield energies, offering new ways to understand and potentially manipulate these fields for therapeutic purposes.

"Foundations of Energy Medicine" sets the stage for a deeper exploration of how energy medicine practices, inspired by both Tesla's work and ancient healing traditions, leverage the human energy anatomy to foster wellness and healing. It invites readers to consider the body not just as a physical system but as a complex interplay of energies, opening up new paradigms for understanding health and disease.

The Role of Water in Energy Medicine: Exploring Water's Conductivity

Water's fundamental role in the human body transcends its biological necessity; it is also a pivotal conductor of bioelectrical signals, playing a crucial role in the transmission of energy and information within the body. This section explores water's unique properties that make it an indispensable element in the context of energy medicine, bridging the gap between physical health and energetic healing.

WATER AS A CONDUCTOR OF BIOELECTRICAL SIGNALS

The human body, composed of approximately 60% water, relies on this vital fluid for more than just the sustenance of life processes. Water's exceptional conductivity allows it to efficiently transport ions and charged particles, facilitating the smooth operation of the body's bioelectrical system. This system is essential for everything from neural communication and muscle contraction to the regulation of the heart and metabolic processes. In the realm of energy medicine, understanding water's conductive properties opens avenues for exploring how energetic therapies might influence physiological functions.

WATER'S ROLE IN TRANSMITTING ENERGY AND INFORMATION

Beyond mere conductivity, water is thought to play a dynamic role in the storage and transmission of information within the body. Theories suggest that water molecules can form structured arrangements, potentially influenced by electromagnetic fields, vibrational energy, and even intentions and emotions. This structured water, sometimes referred to as "exclusion zone" (EZ) water, could act as a medium for transmitting energetic information across the body, influencing health and disease patterns.

MASARU EMOTO'S WORK ON THE ENERGETIC IMPRINTING OF WATER

Japanese researcher Masaru Emoto brought worldwide attention to the idea that water could be energetically imprinted with thoughts, words, and emotions. Through his experiments, Emoto demonstrated that water exposed to positive speech and thoughts formed aesthetically pleasing ice crystals, whereas water exposed to negative intentions did not. Although Emoto's work is controversial and lacks rigorous scientific validation, it has significantly influenced the field of energy medicine by suggesting that water's role in the body could extend beyond the physical to the energetic and informational realms.

IMPLICATIONS FOR HEALTH AND HEALING

The concept of water as a conductor and transmitter of bioelectrical signals and energetic information has profound implications for health and healing. It suggests that therapies aimed at modifying the body's energetic fields, such as Reiki, acupuncture, and even sound therapy, might exert their effects in part through the medium of water. Understanding and harnessing water's conductive and informational properties could lead to novel approaches in energy medicine, potentially offering new pathways for promoting health, healing, and well-being.

In conclusion, water's role in energy medicine is a fascinating area of exploration that bridges the scientific and the speculative, offering insights into how the most basic element of life might be integral to the complex interplay of energy and health. As research continues to unveil the mysteries of water's properties, the potential for innovative healing modalities rooted in the understanding of water's conductivity and informational capacity appears ever more promising.

The Impact of Electromagnetic Pollution: Navigating Electromagnetic Environments

In the modern world, the air around us is invisibly teeming with electromagnetic fields (EMFs) generated by a multitude of sources, from the cell phones in our pockets to the Wi-Fi routers that connect us to the global network. While these technological advancements have undoubtedly transformed society, they have also given rise to concerns about electromagnetic pollution and its potential impact on human biofields and overall health. This section delves into the complexities of navigating our electromagnetic environments and offers insights into mitigating the negative impacts, drawing inspiration from Nikola Tesla's vision of harmonizing with natural energies.

UNDERSTANDING ELECTROMAGNETIC POLLUTION

Electromagnetic pollution, or electrosmog, refers to the pervasive exposure to artificial EMFs generated by electronic devices and infrastructure. Unlike the natural electromagnetic fields that have enveloped the Earth since its formation—fields to which life has adapted over millennia—artificial EMFs are a recent introduction to our environment. The rapid proliferation of technology has led to an exponential increase in exposure levels, raising concerns about the long-term effects on biological systems, including the human biofield, which is thought to be sensitive to electromagnetic fluctuations.

EFFECTS ON HUMAN BIOFIELDS AND HEALTH

The human biofield, a concept embraced by energy medicine, refers to the complex electromagnetic fields produced by the body's biological processes. Research into the impact of artificial EMFs on human health is ongoing, but there is evidence to suggest that prolonged exposure may disrupt the natural functioning of the biofield, potentially leading to adverse health outcomes. Reported effects include sleep disturbances, increased stress levels, and a possible increased risk of certain types of cancer, although the scientific community continues to debate the extent and mechanism of these impacts.

STRATEGIES FOR MITIGATING NEGATIVE IMPACTS

Aligning with Tesla's vision of living in harmony with natural energies, there are several strategies individuals can employ to mitigate the potential negative effects of electromagnetic pollution:

Reduce Exposure: Limit time spent on mobile devices and maintain a safe distance from Wi-Fi routers, especially during sleep. Consider using wired connections where possible.

Create Low-EMF Spaces: Designate areas in your home, such as bedrooms, as low-EMF zones by minimizing the use of electronic devices and employing EMF shielding techniques and materials.

EMF Shielding and Protection: Explore the use of EMF shielding devices and materials that can block or reduce exposure to artificial EMFs. This includes special cases for mobile devices, shielding fabrics, and paints.

Grounding and Earthing: Engage in grounding practices, such as walking barefoot on the Earth, to neutralize the body's electrical charge and reconnect with the Earth's natural electromagnetic field.

Wellness and Resilience: Strengthen your biofield and overall health through holistic wellness practices like yoga, meditation, and a balanced diet, enhancing your resilience against potential EMF disruption.

EMBRACING TESLA'S VISION

In navigating the electromagnetic challenges of modern life, we can draw inspiration from Nikola Tesla's work and his profound respect for the natural world. By adopting practices that reduce our exposure to artificial EMFs and strengthen our connection to natural energies, we move closer to Tesla's ideal of living in harmony with the Earth's rhythms and energies. As research into the effects of electromagnetic pollution advances, it will be crucial to continue exploring ways to balance technological progress with the health of our biofields and the planet.

Energy Medicine in Non-Human Systems: Beyond the Human Body

The exploration of energy medicine often focuses on its applications and implications for human health, but the principles underlying this field have broader relevance, extending into non-human systems such as plant biology and ecological systems. This expansion aligns with Nikola Tesla's visionary perspective on energy, underscoring its universal role in the vitality and interconnectedness of all life forms. This section delves into how energy medicine principles apply beyond the human body, illustrating the universality of energetic principles and Tesla's expansive vision of energy's role in life.

ENERGY DYNAMICS IN PLANT BIOLOGY

Plants, like all living organisms, are vibrant energy systems. The concept of energy medicine becomes fascinating when applied to plant biology, particularly in understanding how electromagnetic fields (EMFs) and specific vibrational frequencies can influence plant growth, health, and resilience. Research has shown that plants are sensitive to subtle energy variations in their environment, including exposure to different light frequencies, sound vibrations, and EMFs. For instance, certain frequencies can accelerate seed germination or enhance plant growth and nutrient content, offering insights into how energy management could revolutionize agricultural practices. This sensitivity reflects a form of communication and interaction with the environment that echoes the principles of energy medicine, suggesting ways to harmonize agricultural practices with natural energetic patterns for sustainable and life-enhancing farming.

ECOLOGICAL SYSTEMS AND ENERGETIC INTERCONNECTEDNESS

The application of energy medicine principles extends to entire ecological systems, where the interplay of energy fields among various life forms and the environment plays a crucial role in maintaining ecological balance. Tesla's vision of energy's role in life emphasized not only its utility in technological advancements but also its fundamental importance in the natural world's harmony and vitality. Understanding the energetic interactions within ecosystems can inform conservation efforts, revealing how human-made electromagnetic pollution or alterations to the Earth's natural EMFs can disrupt wildlife behavior, plant growth, and overall ecosystem health. By applying principles of energy medicine at an ecological level, there's potential to develop more sustainable interactions with our environment that respect the energetic integrity of natural systems.

TESLA'S BROADER VISION FOR ENERGY'S ROLE IN LIFE

Tesla recognized that energy is the fundamental building block of the universe, animating the inanimate and connecting disparate forms of life in a web of interaction. His work on wireless energy transmission and his speculative ideas about the Earth itself as a conductor of resonant energy frequencies were not merely technological pursuits; they were part of a broader vision that saw energy as the key to understanding and enhancing life in all its forms. By examining energy

medicine's application in non-human systems, we embrace Tesla's holistic view of energy, recognizing its potential to foster health, growth, and equilibrium beyond the human sphere to the entirety of the natural world.

In conclusion, exploring the concept of energy medicine in non-human systems reveals the deep interconnectedness of all life through the medium of energy. It invites us to consider Tesla's legacy not only in terms of his contributions to electrical engineering and physics but also as a profound inquiry into the essence of life itself. By applying energetic principles across the spectrum of living systems, we open new avenues for enhancing the well-being of our planet and all its inhabitants, truly embodying Tesla's vision of harmonizing with the natural energies that pervade the universe.

Personal Energy Management: Practical Energy Hygiene

In the realm of energy medicine, understanding the theoretical foundations is just the beginning; applying these concepts to manage one's personal energy fields is where the transformative potential truly lies. This section introduces readers to the concept of personal energy management, emphasizing practical energy hygiene practices. These techniques and daily habits are designed to help individuals maintain a balanced biofield, enhancing overall well-being and complementing the theoretical knowledge explored earlier in the book.

GROUNDING TECHNIQUES

Grounding, or earthing, involves direct contact with the earth's surface to balance the body's bioelectrical environment. This simple yet profound practice helps neutralize excess positive charges—believed to accumulate in the body from exposure to electromagnetic pollution and modern lifestyles—and aligns your biofield with the earth's natural energy. Techniques include:

Walking Barefoot: Spend time walking barefoot on natural surfaces like grass, soil, or sand.

Grounding Mats: Use grounding mats or sheets that connect to the earth's electrical field, providing grounding benefits indoors.

Gardening: Engaging in gardening activities, where hands and feet are in direct contact with the earth, offers grounding as well as a therapeutic connection to nature.

ENERGY CLEANSING PRACTICES

Energy cleansing practices are designed to clear the biofield of accumulated negative or stagnant energies, promoting emotional, mental, and physical health. Techniques include:

Salt Baths: Soaking in baths with sea salt or Epsom salt is believed to help cleanse the aura and restore vitality.

Smudging: The practice of burning sacred herbs, such as sage or palo santo, and allowing the smoke to pass around and over the body is used for clearing negative energies.

Visualization: Visualizing a white or golden light enveloping the body can be a powerful tool for cleansing and protecting the energy field.

DAILY HABITS FOR BIOFIELD BALANCE

Incorporating simple daily habits can significantly contribute to maintaining a balanced biofield and enhancing energy hygiene:

Mindful Breathing: Regularly practicing deep, mindful breathing helps regulate the body's energy flow and reduces stress levels.

Healthy Diet: Consuming foods that are high in life force energy, such as fresh fruits and vegetables, supports the vitality of the biofield.

Digital Detox: Taking regular breaks from electronic devices reduces exposure to artificial EMFs and supports biofield health.

Positive Thinking: Cultivating a habit of positive thinking and gratitude can elevate your vibrational frequency and strengthen your biofield.

Conclusion

Personal energy management is an essential aspect of applying energy medicine principles to everyday life. By engaging in grounding techniques, energy cleansing practices, and daily habits for biofield balance, individuals can enhance their well-being on multiple levels. This practical approach to energy hygiene empowers readers to take active steps toward harmonizing their energy fields, bringing the theoretical aspects of energy medicine into the tangible realm of daily living. Through these practices, we embody Tesla's vision of harmonizing with natural energies, promoting health, and enhancing our connection to the universe's dynamic energy web.

"THE DAY SCIENCE BEGINS TO STUDY NON-PHYSICAL PHENOMENA, IT WILL MAKE MORE PROGRESS IN ONE DECADE THAN IN ALL THE PREVIOUS CENTURIES OF ITS EXISTENCE."
—NIKOLA TESLA

Chapter 3

Quantum Healing: Bridging Science and Spirit

In the exploration of energy medicine, one of the most compelling intersections occurs between the realms of quantum physics and holistic healing practices. Chapter 3, "Quantum Healing: Bridging Science and Spirit," delves into this fascinating confluence, uncovering how the principles of quantum mechanics not only reshape our understanding of the physical world but also offer profound insights into the mechanisms of healing at the most fundamental levels of existence.

THE QUANTUM CONNECTION

Quantum physics, the study of the behavior of matter and energy at the smallest scales, reveals a universe far more interconnected and dynamic than previously conceived. At the heart of quantum theory is the understanding that particles can exist in multiple states simultaneously and that observation itself can affect the outcome of events— a principle known as the observer effect. This notion of interconnectedness and the impact of consciousness on physical reality lay the groundwork for quantum healing, a field that suggests the mind's potential to influence health outcomes through quantum phenomena.

The quantum connection in healing explores concepts like entanglement, where particles remain connected such that the state of one (no matter the distance) can instantaneously influence the state of another. This has led to speculations about the interconnectedness of all life and the potential for healing interventions that operate beyond the constraints of time and space. Additionally, the concept of quantum superposition—the ability of a particle to exist in multiple states until measured—parallels the holistic understanding of health as a dynamic balance of potential states of well-being.

Enhancing our understanding of The Quantum Connection within the realm of quantum healing necessitates diving deeper into the foundational principles of quantum physics and their profound implications for the concept of health and healing.

Quantum physics unveils a world where the very fabric of reality is woven from the probabilities and potentialities of quantum particles. This microscopic universe operates under rules that defy classical physics' predictability and determinism. The observer effect, a cornerstone of quantum theory, suggests that the act of observation can fundamentally alter the state of a quantum system. This principle hints at a universe responsive to consciousness, where the mere focus of attention might collapse potentialities into reality. In the context of healing, this raises the tantalizing possibility that our intentions, beliefs, and focused thoughts could directly influence our physical state and health outcomes.

The phenomenon of quantum entanglement further expands our understanding of interconnectedness. Once particles become entangled, they share a state where the action performed on one instantaneously affects the other, regardless of the distance separating them. This phenomenon has profound implications for conceptualizing healing practices that transcend spatial limitations, suggesting that our thoughts and intentions could have direct effects on others' well-being, mediated by a quantum field that connects all matter and energy in the universe.

Quantum superposition introduces another layer of complexity, presenting a reality where particles exist in a state of potentiality, embodying all possible states until observed. This principle mirrors the holistic view of health as a spectrum of potential well-being states, where disease and health are not fixed but fluctuate across a continuum of possibilities. Quantum healing, in this light, is an endeavor to influence the wave function of well-being, encouraging the collapse of quantum states into optimal health through conscious intention and vibrational alignment.

Moreover, the quantum vacuum—the quantum field's baseline energy state, even in the absence of any matter—offers insights into the energetic substratum of existence. This vacuum is not empty but teems with fluctuations and virtual particles, providing a canvas on which the dance of existence unfolds. Healing interventions might tap into this fundamental energy, leveraging the quantum vacuum's potential to influence the physical and energetic aspects of health.

The exploration of wave-particle duality adds another dimension to our quantum understanding. The ability of quantum entities to exhibit both particle-like and wave-like properties suggests a fluidity of being that transcends classical categorizations. In healing, this duality encourages a view of the self and health as dynamic interplays of energy and matter influenced by observation, intention, and environmental interaction.

In sum, The Quantum Connection in healing invites us to reconceptualize health and healing through the lens of quantum physics, where consciousness, intention, and energy interact within an intricately connected universe. It

beckons us toward a model of healing that is as much about aligning with the fundamental energies of the universe as it is about the physical interventions we may employ. This expanded understanding not only aligns with Tesla's vision of harnessing universal energies but also opens new frontiers in our quest for health and wellness, grounded in the profound interconnectedness and potentialities that quantum physics reveals.

From Theory to Practice: Quantum Applications in Healing

Translating the esoteric principles of quantum physics into practical applications in healing requires a leap from theory to practice. This section explores various ways in which quantum healing manifests, offering a bridge between the scientific understanding of quantum mechanics and the intuitive practices of energy medicine:

Intention and Healing: Drawing from the observer effect, quantum healing posits that focused intention can influence health outcomes. Practices like distant healing and prayer, examined through the lens of quantum entanglement, suggest that healing intentions can have effects across distances, potentially mediated by the quantum field.

Visualization Techniques: Quantum healing leverages visualization techniques to harness the mind's ability to affect physical health. By envisioning the body in a state of perfect health, practitioners tap into the quantum potential for well-being, influencing the body's energy field and promoting healing.

Energy Psychology: Modalities like Emotional Freedom Techniques (EFT) and Thought Field Therapy (TFT) incorporate the understanding of the body's energy system within a quantum framework, using tapping on meridian points to address psychological and physical issues, suggesting that shifts in the body's energy field can lead to tangible health improvements.

Quantum Biofeedback: Emerging technologies in quantum biofeedback aim to measure the body's energetic responses to various stimuli, providing real-time feedback that can be used to promote balance and healing. These devices, rooted in the principles of quantum physics, offer a sophisticated means of assessing and influencing the body's energy systems.

Homeopathic Medicine represents a quantum-inspired approach to healing that operates on the principle of "like cures like." At its core, homeopathy involves using highly diluted substances that, if given in higher doses to a healthy person, would produce symptoms similar to those being treated. From a quantum perspective, it's posited that the water in these dilutions retains a memory or imprint of the original substance, affecting the body's quantum field and stimulating a healing response. This notion resonates with the quantum concept of non-locality and the capacity of energy fields to carry information, suggesting that even the subtlest interventions can initiate profound healing processes.

Quantum Touch and Healing Hands techniques exemplify the application of life force energy, or qi, for healing purposes, guided by the healer's focused intention and deep state of heart-centered awareness. Practitioners use techniques to amplify and direct energy, often through their hands, to areas of the body requiring healing. This practice draws upon quantum entanglement and the observer effect, illustrating how directed consciousness and energy can influence physical health, aligning bodily systems to their optimal state of function.

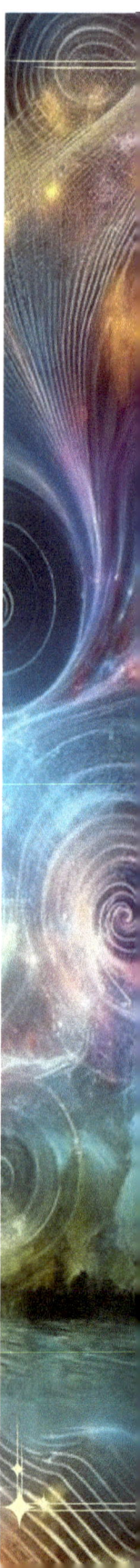

Vibrational Medicine explores healing through the modulation of energy frequencies within the body. This approach includes the use of sound frequencies, light therapy, and even the energetic properties of crystals to harmonize the body's vibrations. The underlying principle aligns with quantum physics' revelation that all matter is fundamentally vibrational. By adjusting these vibrations to resonate with health-promoting frequencies, vibrational medicine seeks to restore balance and promote healing at the quantum level.

Building on the understanding of water's crucial role in biological systems and its capacity to hold and transmit information, **Coherent Water Structuring for Healing** delves into the manipulation of water's molecular structure to enhance its coherence and healing potential. Techniques may involve exposure to specific electromagnetic frequencies, structured light, or sound vibrations that organize water molecules into more ordered states. This structured, or coherent, water is thought to be more efficient in conducting bioelectrical signals and facilitating healing processes, reflecting Tesla's insights into the importance of resonance and energy harmonization.

Chapter 3 concludes with a reflection on the implications of quantum healing for the future of medicine. By bridging the gap between the scientific and the spiritual, quantum healing challenges the conventional boundaries of health and wellness, inviting a deeper exploration of the interconnectedness of mind, body, and spirit. As we venture further into the quantum realm, the potential for transformative healing practices that harness the fundamental energies of the universe becomes ever more apparent, echoing Tesla's visionary insights into the energetic fabric of reality.

"EVERYTHING IS ENERGY, AND THAT IS ALL THERE IS TO IT. MATCH THE FREQUENCY OF THE REALITY YOU WANT, AND YOU CANNOT HELP BUT GET THAT REALITY. IT CAN BE NO OTHER WAY. IT IS NOT PHILOSOPHYIT IS PHYSICS."
—ALBERT EINSTEIN

Chapter 4

The Power of Intention and Consciousness

Chapter 4 delves into the transformative realm where intention and consciousness converge with the physical world, showcasing their profound implications for healing and wellness. This exploration is grounded in the evolving understanding of how our mental states—our focused intentions and conscious awareness—can significantly impact physical reality, a concept that resonates deeply with both ancient wisdom traditions and cutting-edge scientific research.

MIND OVER MATTER: THE SCIENCE OF INTENTION

The adage "mind over matter" encapsulates the essence of how focused intention can influence physical outcomes. Groundbreaking experiments in the field of psychokinesis, the ability of the mind to affect physical objects or events, have provided empirical evidence supporting the power of intention. Moreover, studies in the realm of the placebo effect, where positive health outcomes occur due to the belief in the efficacy of a treatment rather than the treatment itself, underscore the tangible effects of intention and belief on the body.

The principle of "mind over matter," suggesting that the power of the human mind can influence physical reality, is a cornerstone concept in understanding how focused intention can impact physical outcomes. This idea, deeply rooted in various philosophical and spiritual traditions, has gained empirical support from scientific investigations in areas such as psychokinesis and the placebo effect.

Psychokinesis: The Mind's Influence Over Matter

Psychokinesis (PK) represents the purported ability of the mind to influence or move physical objects without physical interaction. While often associated with paranormal phenomena, psychokinesis has been the subject of scientific investigation. Researchers like J.B. Rhine and others have conducted controlled experiments attempting to measure the influence of mental intention on physical systems, such as random number generators or the behavior of objects in sealed containers. Although results have varied and the field remains controversial, some studies report statistical deviations from expected outcomes, suggesting that the focused intention may indeed have an observable effect on physical matter. These studies open intriguing questions about the nature of consciousness and its interaction with the physical world.

The Placebo Effect: Belief and Healing

The placebo effect is a phenomenon where patients experience real improvements in their health after receiving a treatment that has no therapeutic value solely based on their belief in the treatment's efficacy. This effect has been extensively documented across numerous clinical trials, demonstrating that the expectation of healing can trigger significant physiological changes, ranging from pain relief to the modulation of immune response. The placebo effect underscores the powerful role of belief and intention in the healing process,

suggesting that the mind's expectations can influence the body's biology.

For instance, in studies where patients were given sugar pills instead of actual pain medication and told they were receiving a potent analgesic, many reported substantial reductions in their pain levels. Neuroimaging studies have even shown that the placebo effect can activate the same brain areas involved in pain perception, releasing natural pain-relieving chemicals such as endorphins.

Integrating Mind and Matter

These areas of research—psychokinesis and the placebo effect—though distinct, both highlight the profound potential of the human mind to influence physical outcomes. They challenge the conventional materialistic view of health and medicine, suggesting a more integrated approach where mental states, beliefs, and intentions are recognized as potent factors in the healing process. This aligns with the broader principles of energy medicine, which advocate for a holistic view of health encompassing body, mind, and spirit.

The exploration of "mind over matter" through psychokinesis and the placebo effect offers a fascinating glimpse into the untapped potential of focused intention and belief in influencing physical reality and health outcomes. These phenomena invite us to consider a more expansive view of healing, where the power of the mind is harnessed as a fundamental component of wellness and medical care.

This section also explores research conducted in the field of distant healing and prayer, demonstrating that the intentions of a healer can positively affect the health of a patient, even when they are geographically separated, pointing towards a non-local aspect of consciousness that transcends space and time. Additionally, the work of scientists like Dean Radin and experiments like the double-slit experiment under controlled conditions reveal how the mere act of observation and intention can alter the behavior of particles at the quantum level, further evidencing the profound connection between mind and matter.

The exploration of distant healing and prayer within the context of energy medicine brings to light the remarkable concept that consciousness—and the intentions it generates—can transcend the limitations of space and time to effect change. This notion not only challenges the traditional paradigms of medicine and healing but also aligns with the principles of quantum physics, suggesting a deeper, more interconnected fabric of reality.

Distant Healing and Prayer

Distant healing and intercessory prayer involve directing healing intentions or prayers toward individuals in need, regardless of their physical location. This practice is based on the belief in a non-local aspect of consciousness that enables a healer's intentions to influence a patient's health from a distance. Numerous studies have attempted to quantify the effects of distant healing on various health outcomes, with mixed results. However, some research has shown statistically significant effects on recovery rates, pain reduction, and emotional well-being, suggesting that the focused intent of one individual can positively affect another, even over long distances.

The implications of these findings point toward an understanding of consciousness as a non-local field that can influence physical reality, a perspective that resonates with many spiritual and philosophical traditions worldwide. It also raises intriguing questions about the mechanisms through which these effects are mediated.

Quantum Observations: The Work of Dean Radin and the Double-Slit Experiment

Dean Radin, among other scientists exploring the boundaries of consciousness and matter, has conducted experiments investigating the role of observation and intention at the quantum level. Radin's experiments, building on the foundational double-slit experiment in quantum physics, have sought to determine whether conscious intention can influence the behavior of particles.

The double-slit experiment famously demonstrates that particles such as electrons and photons can display characteristics of both waves and particles, depending on whether they are observed. When not observed, they appear to pass through two slits simultaneously, creating an interference pattern indicative of wave-like behavior. However, when observed, they behave like particles, passing through one slit or the other.

Radin's work extends this concept by introducing the variable of human consciousness, testing whether the focused intention of an observer can influence which path particles take or alter the interference pattern. While these experiments challenge the conventional understanding of observer neutrality in scientific observation, some of Radin's results suggest that consciousness can indeed interact with and influence quantum systems.

Bridging Mind and Matter

The research into distant healing, prayer, and the impact of intention on quantum systems underscores a profound connection between consciousness and the physical world. These findings hint at a universe more interconnected and responsive to the mind than previously thought, suggesting potential applications for healing and wellness that leverage this interconnectedness.

By acknowledging the non-local aspect of consciousness and its ability to influence matter across distances and even at the quantum level, we open new avenues for understanding healing. This perspective not only expands the scope of traditional medicine but also aligns with Nikola Tesla's visionary ideas about energy and the universe's interconnected fabric, inviting a deeper exploration of the ways in which focused intention and consciousness can be harnessed for health and healing.

Schrödinger's Cat: The Quantum Dilemma

Schrödinger's cat is a thought experiment that illustrates the peculiarities of quantum mechanics, especially the concept of superposition. In this experiment, a cat is placed in a sealed box along with a radioactive atom, a Geiger counter, a hammer, and a vial of poison. If the radioactive atom decays, the Geiger counter triggers the hammer to release the poison, killing the cat. According to quantum mechanics, until the box is opened and the cat is observed, the cat is considered to be simultaneously alive and dead, embodying a superposition of states.

The thought experiment was designed to challenge the Copenhagen interpretation of quantum mechanics, which posits that a quantum system remains in superposition until it's observed, at which point it 'collapses' into one state or another. Schrödinger's cat brings to light the paradox of applying quantum mechanical principles to everyday objects, highlighting the difficulties in understanding quantum mechanics' implications for reality at larger scales.

Implications for Consciousness and Reality

While Schrödinger's cat primarily addresses quantum mechanics' interpretations, it also opens broader

philosophical discussions about the nature of reality, observation, and the role of consciousness. The idea that the act of observation can determine the state of a quantum system suggests a fundamental interplay between consciousness and the physical world, resonating with themes explored in quantum healing and the power of intention.

Connecting to Quantum Healing

In the context of quantum healing, the principles highlighted by Schrödinger's cat—superposition and the role of the observer—underscore the potential influence of consciousness on health and reality. The thought experiment serves as a metaphor for the possibilities inherent in the human body and mind, where focused intention and belief may influence health outcomes in seemingly paradoxical ways, echoing the quantum dilemma of the cat's simultaneous states.

David R. Hawkins

David R. Hawkins (1927-2012) was a psychiatrist, spiritual teacher, and author known for his work in psychiatry, consciousness research, and spirituality. Hawkins' work spans a wide range of topics, including health, psychology, and spiritual growth, but he is perhaps most famous for his contributions to understanding consciousness and his development of a "Map of Consciousness."

Map of Consciousness

One of Hawkins' most influential concepts is the Map of Consciousness, presented in his book "Power vs. Force" and further elaborated in subsequent works. This map is a scale that purports to calibrate the levels of human consciousness. The scale ranges from 1 to 1000, with lower levels associated with negative emotions and states (such as shame, guilt, and fear) and higher levels corresponding to positive, spiritually advanced states (like love, joy, and enlightenment). Each level is associated with specific emotions, behaviors, and worldviews.

Hawkins claimed that this map could be used to quantify and categorize the consciousness level of individuals, groups, and even concepts or objects, using a technique he called "applied kinesiology" or muscle testing. He suggested that human consciousness has the power to affect reality and that by raising one's consciousness level, individuals can contribute to positive change in the world.

Applied Kinesiology and Muscle Testing

Hawkins used applied kinesiology, or muscle testing, as a central method for validating the principles outlined in his Map of Consciousness. According to Hawkins, muscle testing could be used to assess the truth or falsehood of statements, the integrity of objects, and the consciousness level of individuals or concepts. He proposed that the body's physical response in a muscle test (strength or weakness) could indicate alignment or misalignment with higher truth and consciousness levels.

Criticism and Controversy

While David R. Hawkins' work has been influential in spiritual and self-help circles, it has also faced criticism and skepticism, particularly from the scientific community. Critics argue that applied kinesiology and muscle testing are not reliable or scientifically validated methods for measuring truth or consciousness. Additionally, the subjective nature of Hawkins' Map of Consciousness and its calibration method have raised questions about the objectivity and reproducibility of his findings.

Legacy

Despite the controversy, Hawkins' work has left a lasting impact on discussions about consciousness, spirituality, and personal development. His Map of Consciousness, in particular, has been embraced by many as a useful framework for understanding personal growth and spiritual evolution. Hawkins' books and teachings continue to inspire those on a path of self-discovery and those seeking to understand the deeper, spiritual aspects of human existence.

David R. Hawkins' work represents a bridge between spirituality and a form of empirical inquiry, inviting individuals to explore the nature of consciousness and its influence on the physical and non-physical realms.

Though Schrödinger's cat and David R. Hawkins' work address different aspects of science and spirituality, both contribute to a broader understanding of the complexities and mysteries surrounding consciousness, reality, and the potential for healing beyond conventional paradigms.

HARNESSING CONSCIOUSNESS FOR HEALING

Expanding on the scientific basis of intention, this part of the chapter discusses practical applications and techniques for harnessing consciousness in the service of healing. It covers meditative practices that cultivate a focused and compassionate state of mind, visualization techniques that leverage the brain's capacity to influence the body, and mindfulness practices that enhance awareness and presence, thereby optimizing the body's innate healing capabilities.

The concept of "healing intentionality" is introduced, which involves directing positive, healing-focused energy toward oneself or others, a practice supported by various healing traditions and now being investigated in clinical settings. Techniques like guided imagery, affirmations, and the intentional use of symbols and rituals in therapeutic settings are examined for their ability to harness the power of intention and consciousness to facilitate healing.

The chapter concludes by considering the broader implications of understanding and utilizing the power of intention and consciousness in healthcare and personal wellness. It suggests a paradigm shift towards an integrative approach to health that acknowledges the critical role of the mind in the healing process. By embracing the power of intention and consciousness, individuals can actively participate in their healing journey, leveraging their inner resources to promote health, resilience, and well-being.

Through "The Power of Intention and Consciousness," readers are invited to explore the invisible yet palpable forces of intention and awareness, unlocking new potentials for healing and personal transformation that align with the holistic vision of interconnectedness and empowerment at the heart of energy medicine.

"Everything is Energy. Your Thought begins it, Your emotions amplifies it, and Your Actions increase its Momentum." —Author Unknown

Chapter 5

Electromagnetic Fields and the Future of Therapy

Chapter 5 ventures into the dynamic interplay between electromagnetic fields (EMFs) and their therapeutic potential, marking a pivotal intersection of science, medicine, and the overarching principles of energy healing. This exploration delves into the foundational understanding of EMFs, their impact on biological systems, and the cutting-edge innovations that harness these fields for health and healing.

UNDERSTANDING ELECTROMAGNETIC FIELDS AND THEIR IMPACT ON HEALTH

Electromagnetic fields (EMFs) are omnipresent in our environment, emitted by both natural and man-made sources. Naturally occurring EMFs, such as the Earth's magnetic field, are essential for the orientation and navigation of migratory species across the globe. Sunlight, another source of natural EMFs, plays a pivotal role in human health, primarily through the synthesis of vitamin D, which is crucial for bone health and immune function.

In contrast, our modern world is saturated with man-made EMFs generated by an array of technological innovations. From the mobile phones in our pockets to the power lines crisscrossing landscapes and the sophisticated medical equipment in hospitals, these

sources emit EMFs across a broad spectrum of frequencies. This spectrum ranges from extremely low frequencies, characteristic of household appliances and electrical grids, to the very high frequencies found in medical imaging technologies like X-rays and gamma rays.

THE BIOLOGICAL INTERACTIONS OF EMFS

The interface between EMFs and biological systems has garnered increasing attention from the scientific community. Research suggests that EMFs, depending on their frequency and level of exposure, can interact with cellular processes in complex ways. For instance, certain frequencies are known to influence ion channels on cell membranes, potentially affecting cellular function and communication.

These interactions raise questions about the implications of EMFs for human health. While the Earth's natural magnetic fields and sunlight have long been recognized as beneficial, the health effects of prolonged exposure to man-made EMFs are subject to ongoing research and debate. Concerns have been raised regarding potential links between EMF exposure and various health issues, including sleep disturbances, neurological effects, and even a possible increase in cancer risk.

THE THERAPEUTIC POTENTIAL OF EMFS

Amidst the concerns, there exists a compelling avenue of research exploring the therapeutic applications of EMFs. This field seeks to harness the ability of controlled EMF exposure to influence biological systems positively. For example, Pulsed Electromagnetic Field (PEMF) Therapy utilizes specific EMF frequencies to stimulate cellular repair, showing promise in accelerating the healing of bone fractures, reducing inflammation, and alleviating pain.

Similarly, innovations such as Transcranial Magnetic Stimulation (TMS) leverage EMFs to target neurological conditions. By applying magnetic fields to specific areas of the brain, TMS has emerged as an effective treatment for major depressive disorder, offering hope to individuals for whom traditional therapies have been ineffective.

SAFETY AND REGULATION

As the therapeutic use of EMFs continues to evolve, so too does the importance of understanding the parameters within which these treatments are both safe and effective. Regulatory bodies and scientific organizations worldwide are engaged in ongoing efforts to establish guidelines that minimize potential risks associated with EMF exposure. These guidelines are critical for ensuring that the benefits of EMF-based therapies are realized without compromising individual health.

The realm of electromagnetic fields represents a fascinating intersection of natural phenomena and technological advancement. As research into the health effects and therapeutic potential of EMFs advances, it is imperative to balance innovation with precaution. By continuing to explore the intricate ways in which EMFs interact with biological systems, we can unlock new possibilities for healing and well-being grounded in a thorough understanding of both the benefits and risks associated with these ubiquitous forces of energy.

Principals of Energy Healing

Energy healing is a broad term that encompasses a variety of practices aimed at manipulating the energy circuits in our bodies to regain balance and facilitate our body's innate healing mechanisms. Despite the diversity of techniques and traditions, several core principles underpin the practice of energy healing, reflecting a holistic approach to health and well-being:

1. The Human Body is an Energetic System

This principle acknowledges that beyond the physical body, humans are composed of energy fields (sometimes referred to as biofields, auras, or chakras) that can influence physical, emotional, and mental health. Energy healing is based on the understanding that disruptions or imbalances in these energy fields can lead to disease or emotional disturbances.

2. Everything is Interconnected

Energy healing operates on the belief in the interconnectedness of all things—our bodies, minds, spirits, and the universe. This principle suggests that health is a reflection of harmony within this interconnected system and that healing involves restoring balance not only within the body but also in the individual's external environment and relationships.

3. The Body has an Inherent Ability to Heal Itself

One of the foundational beliefs of energy healing is that the body possesses a natural ability to heal itself. Energy healing practices aim to support and enhance this self-healing capacity by restoring the flow and balance of energy within the body, thereby facilitating recovery from illness and promoting overall wellness.

4. Mind, Body, and Spirit are Interdependent

Energy healing views health and well-being as the product of harmony between the mind, body, and spirit. It emphasizes the importance of mental and emotional states in physical health and advocates for healing practices that address all levels of the individual—physical, emotional, mental, and spiritual.

5. Healing Can Occur at a Distance

Reflecting the principle of interconnectedness and the non-local nature of energy, many energy healing practices assert that healing energy can be transmitted over distance. This is based on the understanding that space does not restrict the flow of energy and that intention can direct healing energies to individuals, regardless of their physical location.

6. Intention is Powerful

Intention is considered a potent force in energy healing. The focused intention of the healer (and sometimes the recipient) is believed to direct and amplify the healing energy. This principle aligns with the concept that consciousness can influence physical reality, including the health and balance of the energetic system.

7. Healing is a Holistic Process

Energy healing is not just about addressing specific symptoms or diseases but involves treating the whole person. This holistic approach seeks to bring about wellness by restoring balance and harmony in all aspects of an individual's life, recognizing that physical health is deeply connected to emotional, mental, and spiritual well-being.

These principles reflect the depth and breadth of energy healing as a complementary approach to health and wellness. By focusing on the energetic dimensions of health, energy healing practices offer a unique and holistic pathway to healing, emphasizing balance, interconnectedness, and the body's natural capacity for self-recovery.

INNOVATIONS IN ELECTROMAGNETIC THERAPIES

The field of electromagnetic therapies is rich with innovation, leveraging the nuanced understanding of how EMFs interact with the human body to develop treatments that can stimulate healing, relieve pain, and improve health outcomes. Some of the most notable advancements include:

- Pulsed Electromagnetic Field Therapy (PEMF): PEMF devices emit electromagnetic waves at various frequencies to stimulate and encourage the body's natural recovery process. This therapy has shown promise in treating conditions ranging from fractures and wounds to depression and inflammation by enhancing cellular repair and circulation.
- Transcranial Magnetic Stimulation (TMS): TMS is a non-invasive procedure that uses magnetic fields to stimulate nerve cells in the brain. It's particularly noted for its effectiveness in treating major depression and other neurological conditions by enhancing neurotransmitter levels in the brain.
- Bioresonance Therapy: This therapy posits that every cell and organ in the human body emits its own electromagnetic waves. Bioresonance devices measure these frequencies and then send back

corrected frequencies to the body to promote balance and healing. It's used to diagnose and treat various conditions, including allergies, chronic pain, and digestive issues.

- Magnetotherapy: Utilizing static magnetic fields, magnetotherapy involves applying magnetic devices or magnets to the body's surface to relieve pain and improve health conditions. It's often used for orthopedic injuries and for enhancing recovery post-surgery.

The future of electromagnetic therapies lies in the continued exploration of the optimal frequencies, exposure durations, and modalities that harmonize with the body's natural healing mechanisms. By understanding the precise ways in which electromagnetic fields influence cellular and systemic health, researchers and practitioners can refine these therapies, making them more effective and tailored to individual needs.

As we stand on the threshold of new discoveries in the realm of electromagnetic therapies, the integration of these technologies into mainstream healthcare represents a promising frontier in medicine. These innovations not only embody the principles of energy healing, rooted in the understanding of the body as an interconnected energetic system but also reflect a broader shift towards holistic and non-invasive treatment modalities. Chapter 5 encapsulates a vision of the future where electromagnetic therapies play a pivotal role in fostering health and healing, guided by a deepened understanding of the intricate dance between energy fields and the living tapestry of the human body.

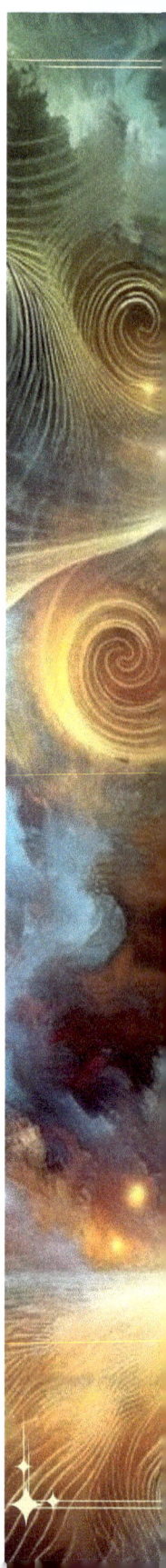

"EVERYTHING IS ENERGY. YOUR THOUGHT BEGINS IT, YOUR EMOTIONS AMPLIFIES IT, AND YOUR ACTIONS INCREASE ITS MOMENTUM."
—AUTHOR UNKNOWN

Chapter 6

The Biofield and Beyond

Chapter 6 delves into the concept of the biofield—a term that encapsulates the complex energy and information fields that are believed to permeate and extend beyond the human body. This chapter explores the foundational understanding of the human biofield, its significance within the context of holistic health, and the diverse array of biofield therapies that aim to manipulate these energy fields for healing and well-being.

THE HUMAN BIOFIELD

Scientifically, the concept of the human biofield refers to a complex of electromagnetic fields and subtle energy systems that are posited to surround and permeate the human body. This concept is rooted in the understanding that all living organisms, including humans, are composed not only of physical matter but also of energy fields that interact with their internal and external environments.

SIMPLIFIED VERSION

Imagine you're playing with a superhero action figure that has a shield of invisible energy all around it. This shield can protect the superhero from harm and even help it heal when it's injured. Well, some people believe that just

like that superhero, we all have our own invisible energy shield called the "human biofield."

Think of the human biofield as a bubble of energy that surrounds and goes through your body. It's not something you can see with your eyes, but it's there, like Wi-Fi signals. This energy bubble isn't just hanging around doing nothing; it's super busy! It helps keep your body working right, like making sure your heart beats and your brain thinks, and it can even affect how you feel— happy, sad, energetic, or tired.

Scientists and healers who study the human biofield think that when this energy bubble is balanced and strong, you feel healthy and full of energy. But if it gets out of balance, like if some parts are too weak or too strong, you might not feel so good. You could get sick more easily or feel stressed or sad.

Some special kinds of doctors and healers try to help people by working with their biofield. They use different ways to help balance this energy, like using their hands to guide the energy (without even touching you!) or teaching you how to imagine or think in ways that can strengthen your biofield. They believe that by helping balance your biofield, they can help you feel better both in your body and your mind.

So, the human biofield is like your very own superhero energy shield, helping to protect you and keep you healthy. And just like superheroes, we can learn to take care of our biofield and keep it strong so we can feel our best!

FOUNDATIONS IN ELECTROMAGNETISM

At the core of the scientific investigation into the human biofield is the recognition that biological processes are fundamentally electromagnetic in nature. The human body generates electromagnetic fields through various physiological activities. For instance:

- Heart and Brain Activity: The heart's rhythmic contractions and the brain's neural activities produce detectable electromagnetic fields, measurable through techniques such as electrocardiography (ECG) and electroencephalography (EEG), respectively.
- Cellular and Molecular Processes: On a smaller scale, cellular processes, including ion transport across cell membranes and mitochondrial energy production, involve electromagnetic phenomena that contribute to the body's overall biofield.

BIOFIELD RESEARCH

The study of the human biofield intersects with several scientific disciplines, including biophysics, bioelectromagnetics, and quantum mechanics. Researchers in these fields explore how biofields interact with physical, chemical, and biological systems, aiming to understand their roles in health and disease. Some areas of focus include:

- Energy Medicine: Investigating therapeutic practices that claim to manipulate the biofield for healing purposes, such as Reiki, Healing Touch, and acupuncture. These practices suggest that interventions in the biofield can influence biological processes and promote health and well-being.
- Biophotonics: The study of biophotons, weak light emitted by living organisms, is another aspect of biofield research. Scientists explore the possibility that biophotons may facilitate

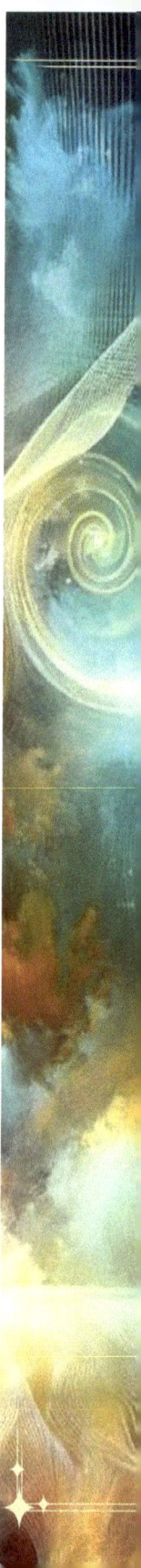

communication between cells and play a role in regulating biological functions.

- Water's Role in the Biofield: Considering the body's high water content, some researchers study how structured water within the body might interact with biofields, potentially influencing cellular activities and health outcomes.

CHALLENGES AND CONTROVERSIES

The scientific study of the human biofield faces several challenges. Measuring subtle energy fields and their effects on health outcomes is complex, requiring highly sensitive instruments and methodologies. Additionally, the biofield concept challenges traditional biomedical paradigms, leading to skepticism and calls for rigorous empirical evidence among the scientific community.

Despite these challenges, interest in the biofield continues to grow, driven by promising research findings and the potential for biofield-based interventions to complement conventional medical treatments. As scientific techniques and instruments become more sophisticated, researchers hope to gain deeper insights into the nature of the biofield, its mechanisms of action, and its implications for health and healing.

In summary, the human biofield, from a scientific perspective, represents an emerging and interdisciplinary field of study that explores the electromagnetic and subtle energy aspects of human physiology and their potential roles in health and disease.

EXPLORING THE HUMAN BIOFIELD

The biofield concept emerges from a synthesis of ancient healing traditions and contemporary scientific investigations, proposing that the human body is not only composed of physical matter but is also enveloped and penetrated by invisible fields of energy and information. These fields are thought to regulate everything from cellular function to our mental, emotional, and spiritual health.

Recent advances in science, particularly within biophysics and quantum mechanics, have begun to provide frameworks for understanding how the biofield might interact with the physical body. Techniques such as gas discharge visualization (GDV) and bioelectromagnetic (BEM) imaging have offered insights into the existence and potential influence of biofields, supporting the notion that health and disease may be significantly influenced by the state of these energy fields.

The exploration of the biofield challenges conventional biomedical paradigms by suggesting that health is not merely the absence of disease but a state of harmonic balance within and among these energy fields. It posits that disturbances or imbalances in the biofield can lead to physical, emotional, and mental health issues, and that restoring balance to the biofield can facilitate the body's natural healing processes.

Dr. Valerie Hunt

Dr. Valerie Hunt was a pioneering researcher, scientist, and professor emerita of Physiological Science at UCLA. Born in 1916, she made significant contributions to the fields of bioenergy and biofields through her innovative research and writings. Dr. Hunt's work spanned several decades, during which she sought to bridge the gap between conventional science and the understanding of human energy fields, also known as biofields or auras.

Key Contributions and Research

Biofield Research

One of Dr. Hunt's most notable contributions was her research on the human biofield. She conducted groundbreaking studies that utilized electromyography (EMG) to measure electrical activity in the muscles and other bioelectrical phenomena associated with the human body. Her research aimed to scientifically validate the existence of the human biofield, which she posited could be detected, measured, and significantly affected health and consciousness.

Dr. Hunt's experiments were among the first to provide empirical evidence suggesting that the human energy field is more complex and contains higher frequencies than those produced by merely physiological activities like heartbeats and brain waves. Her work demonstrated that these energy fields could change in response to different emotional states and healing interventions, providing a scientific basis for understanding how practices such as energy healing might influence health.

The "Infinite Mind" and Consciousness Studies

Dr. Hunt was also deeply interested in the study of consciousness and its relationship with the biofield. In her book "Infinite Mind: Science of the Human Vibrations of Consciousness," she explored the idea that the human mind and consciousness extend beyond the physical brain, interacting with and being influenced by the biofield. She proposed that consciousness could be understood through the lens of energy and vibrations, with implications for healing, personal development, and the expansion of human potential.

One particularly intriguing series of investigations in this domain was carried out by Dr. Valerie Hunt. In her research titled "A Study of Structural Neuromuscular, Energy Field, and Emotional Approaches," she captured the frequencies of low-voltage signals emanating from the body during sessions of Rolfing. To do this, she utilized electrodes made of silver and silver chloride attached to the skin. These recorded wave patterns were then examined using Fourier analysis and sonogram frequency analysis. The findings confirmed that the human energy field is comprised of various color bands, aligning with the chakra system. The outcomes from her study, published in February 1988, demonstrated specific color-to-frequency matches in hertz or cycles per second:

- Blue ranged from 250–275 Hz, with an additional frequency of 1,200 Hz
- Green spanned from 250–475 Hz
- Yellow was observed between 500–700 Hz
- Orange frequencies were between 950–1050 Hz
- Red was found in the range of 1,000–1,200 Hz
- Violet frequencies varied widely, from 1,000–2,000 Hz, with additional bands at 300–400 and 600–800 Hz
- White was measured at frequencies from 1,100–2,000 Hz

These results provide a fascinating insight into the correlation between the human energy field's color bands and the chakras, revealing specific vibrational frequencies associated with each color.

Education and Advocacy

Beyond her research, Dr. Hunt was a passionate educator and advocate for the integration of biofield science into broader understandings of health and wellness. She lectured extensively, offering workshops and seminars that educated people on the importance of the biofield and its potential to transform personal and collective well-being. Dr. Hunt's work inspired many practitioners in the fields of energy medicine, holistic health, and integrative therapies, contributing to the growing interest in and acceptance of these practices.

Legacy

Dr. Valerie Hunt left a lasting legacy in the fields of bioenergy and consciousness research. Her pioneering work paved the way for further scientific exploration of the human biofield and its implications for health, healing, and the nature of consciousness. Despite the skepticism her work sometimes faced from the traditional scientific community, her contributions continue to inspire researchers, healers, and individuals seeking to understand the complex interplay between energy, consciousness, and health.

Dr. Hunt's research into biofields represents an important bridge between conventional scientific methodologies and the experiential realities of holistic health practices, emphasizing the potential of an integrated approach to understanding human physiology and wellness.

Dr. Hiroshi Motoyama

Dr. Hiroshi Motoyama (1925-2015) was a Japanese scientist, researcher, and spiritual visionary. He held Ph.D. degrees in both physiological psychology and philosophy and was the founder of the International Association for Religion and Parapsychology. Dr. Motoyama was deeply rooted in the traditions of Shinto and Yoga and dedicated his life to understanding the mechanisms behind spiritual phenomena through the lens of science.

Key Contributions and Research

AMI Device

One of Dr. Motoyama's significant contributions to the study of biofields was the development of the Apparatus for Measuring the Functions of the Meridians and Corresponding Internal Organs (AMI). The AMI device was designed to measure the bioenergetic condition of the body's meridians and chakras, providing a scientific basis for understanding the flow of vital energy (Qi or Prana) and its impact on health. This invention represented one of the first attempts to quantify the energetic pathways traditionally described in acupuncture and yoga, offering a bridge between these ancient practices and modern scientific inquiry.

Chakra and Kundalini Research

Dr. Motoyama conducted extensive research on chakras and kundalini energy, exploring how these subtle energy centers and the awakening of spiritual energy influence psychological and physiological well-being. His work in this area aimed to provide a scientific framework for

understanding spiritual awakening processes and their tangible effects on the body and mind.

Contributions to Spiritual Science

Beyond his scientific research, Dr. Motoyama was a prolific author, writing extensively on the topics of spirituality, the science of consciousness, and the integration of Eastern and Western approaches to health and well-being. His work emphasized the importance of spiritual development for personal health and the evolution of human consciousness.

Legacy

Dr. Hiroshi Motoyama's legacy lies in his profound contributions to the understanding of the human biofield, chakras, and the scientific exploration of spiritual phenomena. His work continues to inspire researchers, healers, and spiritual seekers alike, serving as a testament to the potential for a deeper integration of science and spirituality in our quest for understanding the full spectrum of human experience.

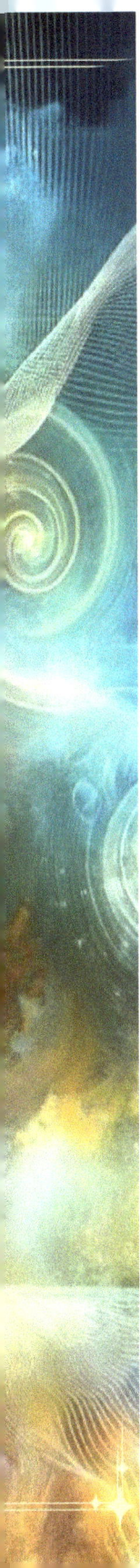

Biofield Therapies: Reiki, Healing Touch, and More

The automatic response of placing your hand on a painful area is a deeply ingrained human instinct that reflects an intuitive understanding of the body's ability to heal and comfort itself. This response can be observed across all ages and cultures, suggesting a universal recognition of touch as a fundamental healing tool.

Scientific Perspectives

Pain Relief and Touch

From a scientific standpoint, this instinctual behavior can be understood through several mechanisms:

- Gate Control Theory of Pain: This theory suggests that non-painful input (such as the pressure from your hand) closes the "gates" to painful input, which prevents pain sensation from traveling to the central nervous system. Simply put, the pressure and warmth from your hand can actually reduce the perception of pain.
- Release of Endorphins: Touch, including self touch, can stimulate the release of endorphins, the body's natural painkillers. These chemicals help alleviate discomfort and promote a sense of well-being.
- Soothing and Emotional Comfort: Beyond the physical mechanisms, touching a painful area can also provide emotional comfort and reassurance. This can trigger a placebo effect, where believing in the action's healing power can itself contribute to pain relief.

The Role of Biofield Therapies

The instinct to touch a place of discomfort aligns with the principles behind many biofield therapies, such as Reiki, Healing Touch, and Therapeutic Touch. These practices are based on the premise that human beings can channel healing energy or manipulate the body's energy field to promote healing and relieve pain. While modern biofield therapies usually involve training and specific techniques, the underlying instinct to use touch as a means of addressing pain is a natural expression of the human body's inherent wisdom and capacity for self-healing.

Evolutionary and Psychological Aspects

Evolutionarily, this instinctual response may have developed as a survival mechanism, providing immediate, albeit temporary, relief from pain and discomfort, which could be crucial in situations where medical help was not readily available. Psychologically, it represents a nurturing gesture towards oneself, signaling care and attention to the body's needs, which can be comforting during moments of pain.

The automatic response of placing one's hand on a painful area is a fascinating intersection of instinct, biology, and learned behavior that illustrates the body's complex mechanisms for coping with discomfort. It underscores the significance of touch as a natural, instinctive tool for healing and comfort, reflecting the body's remarkable ability to initiate self-healing processes.

BIOFIELD THERAPIES

Biofield therapies encompass a wide array of healing practices designed to balance and fortify the human biofield, contributing to overall health and well-being.

These therapies, deeply entrenched in ancient healing traditions from around the globe, utilize a diverse set of techniques to modulate the biofield. This manipulation is predominantly achieved through hands-on healing or the deliberate direction of healing energies.

Hands-on healing itself is a venerable practice, recognized across various cultures under numerous names and forms, each distinguished by its distinct methods and philosophical foundations. Despite their variances, these practices are united by a common purpose: to facilitate healing by purposefully employing the practitioner's hands to influence the biofield. Here are several names and forms of hands-on healing:

Pranic Healing

Originating from the Philippines, this practice involves scanning the body's energy field to detect imbalances and then cleansing and energizing the aura with prana (life energy).

Quantum Touch

A method of natural healing that works with the Life Force Energy of the body to promote optimal wellness. It involves using specific breathing techniques and body awareness meditations.

Bioenergy Healing

Also known as biofield therapy, this broad term encompasses various forms of energy healing where practitioners aim to affect change in the patient's biological energy field.

Chakra Healing

Focusing on the seven chakras, or energy centers, of the body, this form of healing seeks to unblock, balance, and

direct energy throughout the body through hands-on work or energy manipulation techniques.

Polarity Therapy

Developed by Dr. Randolph Stone, this holistic practice combines hands-on contact, diet, exercise, and self-awareness to balance the flow of energy in the body.

Jin Shin Jyutsu

An ancient Japanese art and philosophy for harmonizing the life energy in the body. It involves gentle touch over specific points on the body to restore energy flow.

Craniosacral Therapy

A gentle, hands-on method of evaluating and enhancing the functioning of the physiological body system called the craniosacral system, composed of the membranes and cerebrospinal fluid that surround and protect the brain and spinal cord.

Reflexology

Although often associated with just the feet, reflexology can also be performed on the hands and ears. It involves applying pressure to specific points that correspond to organs and systems in the body.

Shamanic Healing

While not exclusively hands-on, many shamanic healing practices involve direct contact, such as the laying on of hands, to transfer energy, remove energetic blockages, or convey healing intentions.

REIKI

Reiki, a form of energy healing from Japan, operates on the premise that by channeling universal life energy (known as "qi" or "ki") through the practitioner's hands into the recipient, it is possible to promote physical, emotional, and spiritual healing. This process is believed to realign the body's energy flow, remove energy blockages, and restore balance, leading to improved health and well-being. As Reiki has grown in popularity worldwide, its integration into scientific studies and healthcare settings has sparked interest in understanding its effects through a scientific lens.

Scientific Inquiry into Reiki

Physiological Effects

Research into Reiki has sought to quantify its physiological effects, with some studies measuring changes in heart rate, blood pressure, and levels of stress hormones. For example, a controlled study might monitor these physiological markers in patients before and after Reiki sessions, comparing the results with those from a control group receiving standard care or a placebo treatment. Early findings suggest that Reiki may induce a relaxation response akin to the effects of meditation and mindfulness practices, leading to reduced stress and potential improvements in related health conditions.

Clinical Settings Integration

Reiki's integration into clinical settings, particularly in hospitals and palliative care, has provided opportunities for case studies and observational research. Patients undergoing surgery, cancer treatment, or hospice care have reported beneficial effects from Reiki sessions, including reduced pain, anxiety, and nausea, alongside improved mood and overall quality of life. These

anecdotal reports have spurred further investigation into how Reiki might complement conventional medical treatments by supporting patient well-being and recovery.

Case Studies and Anecdotal Reports

- Chronic Pain Relief: There have been individual case studies where patients suffering from chronic pain reported significant relief following Reiki sessions. In some instances, these patients had tried various other treatments with limited success before finding notable improvement through Reiki, suggesting a potential role for energy healing in pain management.
- Post-Surgical Recovery: Anecdotal evidence and some small-scale studies have reported that Reiki can aid in recovery after surgery. Patients have described reduced pain, decreased anxiety, and faster healing times than expected, contributing to an overall sense of well-being and accelerated return to normal activities.
- Cancer Support: Among the most compelling anecdotal reports are those from cancer patients who have incorporated Reiki into their overall treatment plans. Some individuals have credited Reiki with significant reductions in side effects from chemotherapy, such as nausea and fatigue, as well as improvements in emotional resilience and quality of life.
- Stress and Anxiety Reduction: Numerous personal accounts highlight the effectiveness of Reiki in managing stress and anxiety. Patients who have received Reiki treatments often report a profound sense of peace and relaxation, which, in some cases, has led to breakthroughs in dealing with chronic stress, anxiety disorders, and depression.

Scientific Scrutiny and Skepticism

It's important to note that while these case studies and reports are valuable for understanding the potential outcomes of Reiki therapy, they do not constitute scientific proof of its efficacy. The anecdotal nature of these reports and the lack of control groups in many case studies mean that definitive conclusions about Reiki's effectiveness and its ability to produce "miraculous" results cannot be made solely on their basis.

Moreover, skepticism remains in the scientific community regarding the ability of Reiki to produce outcomes beyond those explainable by placebo effects or the natural progression of diseases. Rigorous, controlled studies are necessary to establish a clear scientific basis for any health-related claims.

Challenges and Considerations

One of the main challenges in scientifically studying Reiki is the difficulty in designing rigorous, double-blind studies due to the nature of the practice. The placebo effect—where individuals experience benefits because they believe they are receiving treatment—can complicate the interpretation of results. However, some researchers advocate for the inclusion of "sham" Reiki sessions (where the practitioner does not intentionally channel energy) as a control to help discern the specific effects of Reiki energy healing from placebo responses.

Moreover, the subjective experience of receiving Reiki and the individual differences in response highlight the need for a personalized approach to evaluating its effectiveness. The non-physical nature of "qi" or "ki" presents another scientific challenge, as it eludes direct measurement with current technologies.

As interest in holistic and complementary therapies continues to grow, the scientific exploration of Reiki and

its effects provides a fascinating intersection between ancient healing traditions and contemporary healthcare. Ongoing research, including well-designed clinical trials and comprehensive case studies, is essential for elucidating the mechanisms behind Reiki, its potential health benefits, and its place within an integrative approach to health and healing. Through such inquiry, science seeks to broaden its understanding of healing, acknowledging the complex interplay of mind, body, and energy in promoting health and well-being.

While "miracles" might not be the term commonly used in scientific literature due to its implications of unexplainable phenomena, there are indeed numerous documented case studies and anecdotal reports within the realm of Reiki practice that describe significant, sometimes unexpected, positive outcomes attributed to Reiki treatments. These outcomes often pertain to substantial improvements in physical, emotional, and psychological conditions that were not fully addressed by conventional medical treatments.

HEALING TOUCH

Healing Touch (HT), as a modality of biofield therapy, has garnered interest within the scientific community for its potential impact on various health conditions. By focusing on manipulating the energy field surrounding the body, practitioners aim to promote healing and balance. This interest has led to research studies and clinical trials designed to assess the efficacy of Healing Touch in diverse healthcare settings, including hospitals and private practices.

Impact on Science and Research

Clinical Studies and Reviews

A number of clinical studies have evaluated the effectiveness of Healing Touch in treating conditions such as post-operative pain, anxiety, stress, and the side effects of cancer treatments. These studies often measure quantifiable outcomes related to these conditions, such as pain intensity, stress levels, heart rate variability, and quality of life indices, before and after Healing Touch interventions.

One systematic review, for example, compiled research on Healing Touch's impact on patients undergoing cancer treatment, noting improvements in symptoms like fatigue, pain, and anxiety. Such findings suggest that Healing Touch can be a beneficial complementary therapy in oncology settings, potentially enhancing the well-being and recovery experience of cancer patients.

Integration into Healthcare

The scientific exploration of Healing Touch has also led to its integration into various healthcare practices. Many hospitals now include Healing Touch as part of their integrative medicine programs, offering it alongside conventional treatments to support patient care. This integration is partly due to evidence suggesting that Healing Touch can contribute to reduced anxiety and improved patient satisfaction, particularly in pre- and post-operative settings.

Notable Case Studies

Post-Surgical Recovery

One case study involving Healing Touch focused on a patient recovering from surgery who reported significantly reduced pain levels and faster recovery times than anticipated. The patient received Healing Touch

sessions shortly after surgery, which they credited with not only diminishing their pain without the sole reliance on pharmaceuticals but also with providing a profound sense of peace and emotional balance during recovery.

Stress and Anxiety Reduction in Healthcare Workers

Another interesting application of Healing Touch was in a study involving healthcare professionals experiencing high levels of job-related stress and burnout. Participants who received Healing Touch therapy reported noticeable reductions in stress and anxiety, alongside improvements in overall job satisfaction and resilience. This case highlights the potential for Healing Touch to support not only patients but also healthcare providers.

Challenges and Future Directions

While Healing Touch has shown promise in various clinical and anecdotal reports, challenges remain in firmly establishing its efficacy through scientific means. Skeptics point to the need for more rigorous, double-blind studies to rule out placebo effects and to better understand the mechanisms through which Healing Touch might exert its reported benefits.

Future research directions may involve larger, more diverse study populations, the use of advanced imaging technologies to observe changes in the biofield, and interdisciplinary approaches that bridge conventional medical science with biofield therapy principles.

The scientific exploration of Healing Touch contributes to a growing body of evidence supporting the therapeutic potential of biofield therapies. By examining Healing Touch's effects on health conditions and its integration into healthcare settings, science is gradually uncovering the ways in which energy-based practices can complement traditional medicine, offering holistic pathways to healing and well-being.

THERAPEUTIC TOUCH

Therapeutic Touch (TT), like Healing Touch, is a biofield therapy grounded in the belief that humans are composed of and surrounded by energy fields. Practitioners of Therapeutic Touch assert that by consciously interacting with a patient's energy field, they can identify and alleviate imbalances, thereby promoting health and healing. The scientific exploration of TT, its effects on various health conditions, and its integration into healthcare settings reflect a growing interest in understanding and validating the practice within a scientific framework.

Impact on Science and Research

Empirical Studies

Research on Therapeutic Touch has sought to assess its efficacy in managing pain, reducing stress, and enhancing the healing process. Empirical studies often focus on measurable outcomes such as pain intensity scales, physiological stress markers (like cortisol levels or heart rate variability), and wound healing rates. For instance, studies involving patients with chronic pain conditions or those undergoing post-surgical recovery have examined TT's potential to alleviate discomfort and facilitate quicker recuperation.

One notable study observed the effects of TT on patients with osteoarthritis of the knee, finding that those who received TT experienced significant reductions in pain and stiffness compared to a control group. Such studies contribute to the body of evidence suggesting TT's potential benefits in pain management and support further investigation into its mechanisms of action.

Reviews and Meta-Analyses

Systematic reviews and meta-analyses have been conducted to compile and evaluate the findings from multiple studies on Therapeutic Touch. While results are mixed, with some reviews calling for higher-quality studies, others acknowledge TT's potential to reduce anxiety and improve well-being, particularly in clinical settings. These comprehensive analyses play a crucial role in identifying research gaps, refining methodologies, and guiding future investigations into TT's therapeutic applications.

Integration into Healthcare

The exploration of TT's potential benefits has led to its integration into various healthcare environments, including hospitals, palliative care settings, and private practices. TT is often offered as a complementary therapy alongside conventional treatments, especially in areas where reducing stress, managing pain, and supporting emotional well-being are integral to patient care.

Challenges and Future Directions

Despite growing interest and anecdotal support for TT's efficacy, the practice faces skepticism, primarily due to challenges in quantifying and scientifically validating the manipulation of energy fields. Critics often cite the need for more rigorously designed studies to eliminate placebo effects and subjective bias. Moreover, understanding the underlying mechanisms by which TT may influence health remains a significant scientific challenge.

Future research may focus on employing advanced imaging and measurement technologies to visualize and quantify changes in the biofield before and after TT sessions. Additionally, interdisciplinary studies that bridge conventional medical science with energy healing principles could offer new insights into TT's place within a holistic health model.

Therapeutic Touch represents an intriguing intersection of traditional healing practices and contemporary scientific inquiry. As research into TT and its effects on health continues to evolve, it has the potential to enrich our understanding of the biofield and its role in health and healing. By navigating the challenges of validating TT within a scientific framework, researchers and practitioners can work toward integrating this and similar biofield therapies into a more inclusive, holistic approach to health and wellness, emphasizing the interconnection of mind, body, and energy in the healing process.

THE ROLE OF CASE STUDIES IN ENERGY HEALING

Despite these challenges, case studies play a crucial role in the field of energy healing. They provide insights into patients' experiences, suggest areas for further research, and offer hope to individuals seeking alternative or complementary treatments. As the scientific investigation of Reiki and other energy healing practices continues to evolve, it is possible that more light will be shed on the mechanisms behind these anecdotal miracles, potentially leading to broader acceptance and integration into holistic healthcare approaches.

Conclusion

These and other biofield therapies, such as Quantum Touch and Polarity Therapy, underscore a holistic approach to health and healing, one that acknowledges the interconnectedness of the physical, emotional, mental, and spiritual dimensions of the human experience.

Chapter 6 invites readers to consider the profound implications of recognizing and working with the biofield as an integral aspect of health and healing. By exploring the human biofield and the therapeutic practices that aim to harmonize these energy fields, we open new pathways to wellness that complement traditional medical approaches. As we continue to expand our understanding of the biofield, we not only deepen our connection to ancient healing wisdom but also pave the way for future innovations in holistic health care.

Chapter 7

Tesla's Technologies Reimagined for Health

Chapter 7 explores the innovative legacy of Nikola Tesla, focusing on how his groundbreaking inventions, particularly those related to wireless energy and the Tesla Coil, have inspired modern approaches to energy medicine. Tesla's visionary work has paved the way for a unique intersection between his technological advancements and the field of holistic health, leading to novel therapeutic modalities that harness electromagnetic fields and frequencies for healing purposes.

FROM WIRELESS ENERGY TO HEALING FREQUENCIES

Nikola Tesla's pioneering research into wireless energy transmission has found new resonance in the realm of energy medicine. Tesla envisioned a world interconnected through wireless communication and energy transfer, proposing technologies that could transmit power without the need for physical conduits. Today, this principle of wireless energy transfer is being explored in health and wellness through the development of therapeutic devices that emit specific frequencies intended to stimulate the body's natural healing processes.

The concept of healing frequencies stems from the understanding that every cell and organ in the human

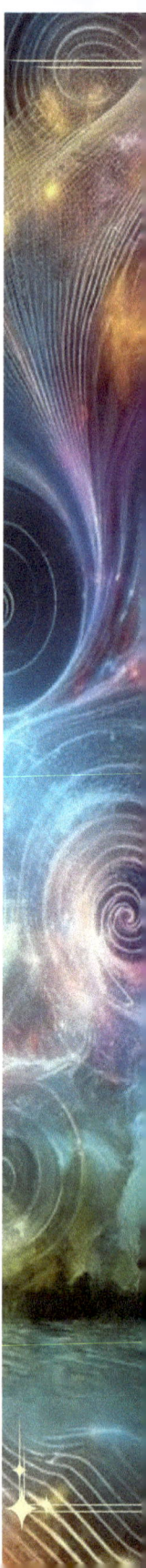

body operates at certain frequencies, and disruptions to these frequencies can lead to disease and imbalance. By applying targeted electromagnetic frequencies, it's theorized that cells can be encouraged to return to their optimal frequency, thereby restoring health. This approach draws directly from Tesla's insights into the electromagnetic spectrum and his experiments with alternating currents and resonant frequencies.

THE TESLA COIL AND MODERN ENERGY MEDICINE DEVICES

The Tesla Coil, one of Tesla's most famous inventions, is a high-voltage, high-frequency power transformer that can generate large electrical fields. While Tesla initially designed the coil for wireless communication and energy transmission, its principles have been adapted for use in various energy medicine devices. These modern devices aim to apply electromagnetic fields to the body in a controlled manner, with the intention of promoting healing, reducing pain, and enhancing cellular function.

For instance, Pulsed Electromagnetic Field (PEMF) therapy devices, which are used to support bone healing, reduce inflammation, and alleviate pain, owe much to Tesla's work with electromagnetic fields. Similarly, devices designed for Transcranial Magnetic Stimulation (TMS) in the treatment of depression and other neurological conditions utilize principles that can be traced back to Tesla's innovations with electromagnetic coils.

TESLA'S INFLUENCE ON ENERGY MEDICINE

Tesla's technological achievements have not only revolutionized the way we use electricity but have also inspired a holistic approach to health that integrates his theories of electromagnetism and resonance. By reimagining Tesla's technologies within the framework of health and wellness, practitioners and innovators in the field of energy medicine are developing therapeutic modalities that offer non-invasive, drug-free alternatives for healing.

The fusion of Tesla's technologies with the principles of energy medicine represents a fascinating blend of science, technology, and holistic health. It underscores the potential of electromagnetic fields and frequencies as tools for healing, opening new avenues for research and application in medical science. As we continue to explore and understand the impact of these technologies on human health, Tesla's legacy remains a beacon of innovation, inspiring future generations to look beyond conventional paradigms in pursuit of holistic wellness solutions.

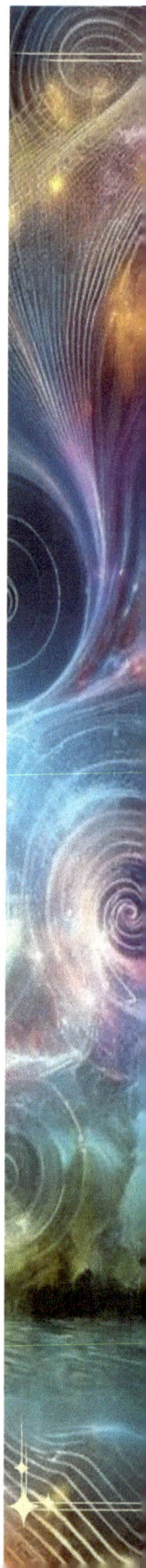

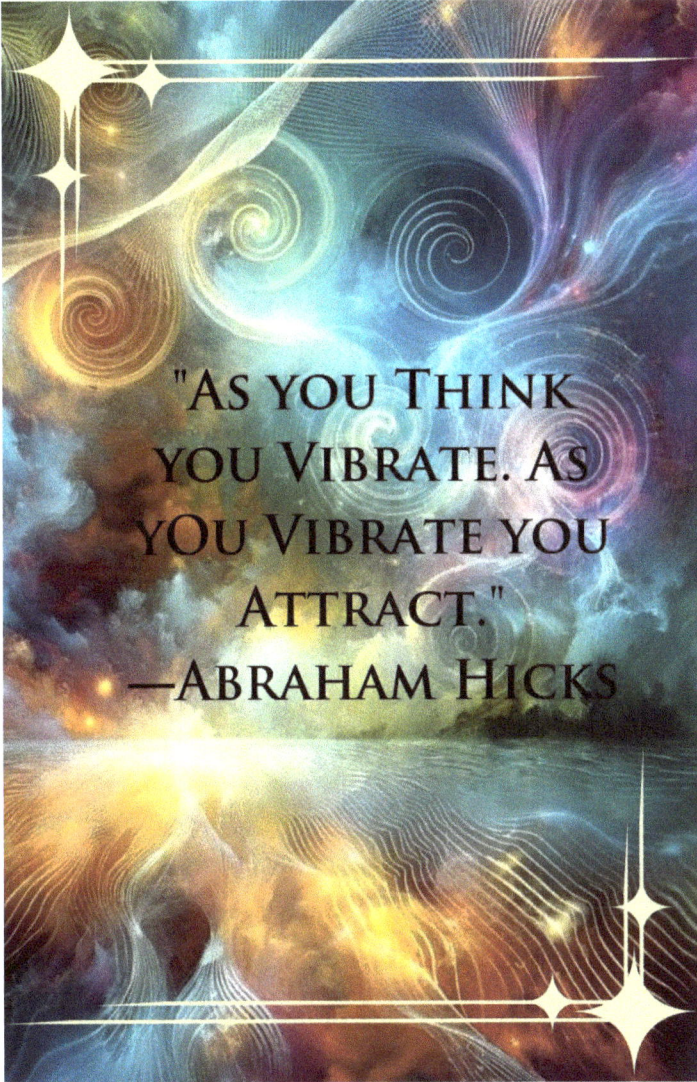

"As you Think you Vibrate. As you Vibrate you Attract."
—Abraham Hicks

Chapter 8

Electromagnetic Spectrum, Frequency, Vibration, and Hertz

ELECTROMAGNETIC SPECTRUM

The electromagnetic (EM) spectrum encompasses all types of electromagnetic radiation. Radiation, in this context, refers to the energy that travels and spreads out as it goes – think of the visible light that comes from a lamp, the radio waves that come from a radio station, or the heat that comes from a hot object. Here's a breakdown of the EM spectrum, starting from the lowest energy/longest wavelength to the highest energy/shortest wavelength:

Gamma-ray   X-ray   Ultraviolet   Visible   Infrared   Microwave   Radio

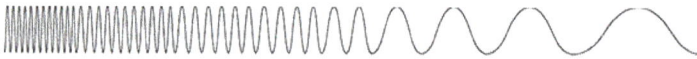

Radio Waves

These are at the lowest energy end of the spectrum, with wavelengths that can be longer than a mountain or as short as a few millimeters. They are used in radio and television broadcasting, wireless communications, and radar.

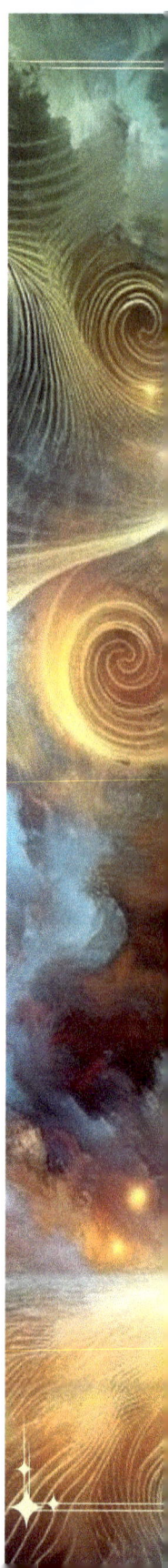

Microwaves

These are similar to radio waves but with shorter wavelengths. They are used in microwave ovens, certain communication transmissions, and for radar. Microwaves are also used by astronomers to learn about the structure of nearby galaxies.

Infrared Radiation

Infrared is primarily heat or thermal radiation. It is used in heaters, for night-vision equipment, and for remote controls for electronic devices. Infrared is also used by astronomers to observe objects in space that are too cool to emit visible light.

Visible Light

This is the part of the spectrum that humans can see. It ranges from red (longer wavelength) to violet (shorter wavelength). This is the light all around us that comes from the sun or from artificial lighting.

Ultraviolet Radiation

Just past the violet end of the visible light spectrum lies ultraviolet (UV) radiation. It has shorter wavelengths and can be harmful in high doses, causing sunburn or even skin cancer. However, it is also used for sterilization and to help make vitamin D in the skin.

X-Rays

Even shorter in wavelength than UV, X-rays can pass through the human body, which makes them useful for imaging internal structures, such as bones in medical diagnostics.

Gamma Rays

Gamma rays have the shortest wavelength in the EM spectrum and are produced by the hottest and most energetic objects in the universe, such as neutron stars and supernovas. They can also be produced by nuclear reactions and are used in certain types of cancer treatments.

Each type of EM radiation has unique properties and interacts with matter in different ways, but all types of EM radiation propagate as waves. The EM spectrum is continuous and seamless, with no precise boundaries between the types of radiation. The classification is based mainly on how the radiation is produced and how it interacts with matter.

FREQUENCY AND VIBRATION

Frequency and vibration are closely related concepts in physics, and they're both fundamental to understanding how energy moves and changes form. Here's a breakdown of what they mean and how they're different:

Vibration:

- What It Is: Vibration is the mechanical oscillation or movement back and forth around an equilibrium point. It's something you can often feel, like when your phone buzzes in your hand.
- Sensation: Because vibration involves physical motion, it can be felt directly. For example, you might sense the vibration of a guitar string when it's plucked or feel the ground shaking during an earthquake.

Frequency:

What It Is: Frequency refers to the number of times a wave repeats (or cycles) in a second. It's a way of measuring the rate at which vibrations occur.

- Units: The unit for frequency is Hertz (Hz), which quantifies how many cycles (vibrations) happen per second. For example, if a drum beats twice in a second, the frequency of the drum's vibration is 2 Hz.
- Sensation: Frequency itself isn't something you can physically touch or feel. It's a measurement. However, the effects of different frequencies can be experienced. High-frequency sounds can be heard as high-pitched tones, while low-frequency vibrations can be felt as rumbles.

Relationship Between Them:

How They Work Together: Vibration can create waves, such as sound waves, which oscillate at certain frequencies. The frequency of these waves determines how we perceive the vibration (as a high or low sound, for example).

Human Perception: Humans can perceive vibrations through our senses, like touch and hearing, within certain ranges. We can hear sound frequencies typically between 20 Hz and 20,000 Hz. Below and above these frequencies, sound waves exist, but they are not audible to humans—they are called infrasound (below 20 Hz) and ultrasound (above 20,000 Hz).

Feeling Frequencies: Sometimes, we can feel the effect of certain frequencies. For example, the bass in loud music is not just heard but also felt through the body. That's because the vibration is large enough for the body to sense it.

Example

When you throw a pebble into a calm body of water, it creates ripples or waves that spread out in concentric circles from the point where the pebble entered the water. This is a great example to explain the concepts of vibration and frequency.

Vibration:

What Happens: The pebble hitting the water's surface disturbs the equilibrium of the water, creating vibrations. This is similar to the initial pluck of a guitar string that causes it to vibrate.

Frequency:

Observation: The ripples consist of waves that move away from the point where the pebble entered the water. Each wave represents one cycle.

Frequency Measurement: If you were to count how many ripples pass by a certain point per second, you'd be measuring the frequency of the wave created by the pebble. The more ripples that pass by in a second, the higher the frequency.

Real-Life Example:

Imagine you're standing at the edge of a pond, and you throw a pebble into the middle. The moment the pebble strikes the water, it sends out ripples. If the pebble is small and thrown gently, it might produce a low number of gentle ripples with a low frequency (fewer cycles per second). If you throw a larger pebble or toss it with more force, it might generate waves that are more vigorous and have a higher frequency (more cycles per second).

In this example, the ripples in the water are the physical manifestation of the vibration, and the number of ripples that pass in a given time is the frequency. While you can

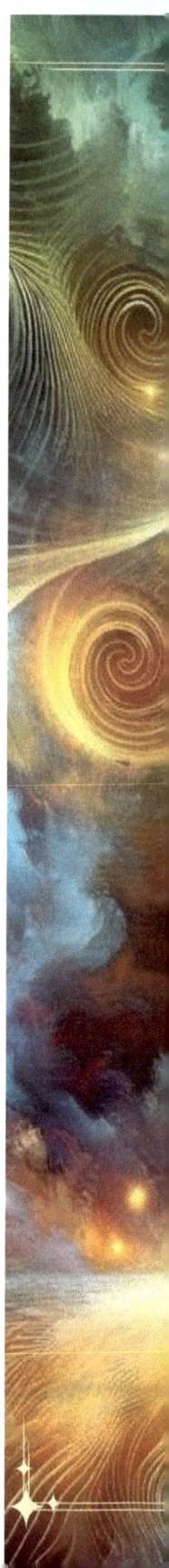

see and perhaps feel (if you're close enough) the ripples, you can't touch or feel the frequency itself, as it's the rate at which the ripples repeat.

In summary, while frequency is a fundamental property of a wave that you can measure, vibration is the actual movement or sensation that you might feel. They are intimately connected, as the frequency of a vibration affects how we perceive it. Understanding these concepts is crucial in various fields, from music to medicine to engineering.

Hertz:

Hertz (Hz) is the unit of measurement for frequency, which is comparable to how centimeters and inches measure length. Named after Heinrich Rudolf Hertz, who was the first to provide definitive proof of electromagnetic waves' existence, the term hertz signifies the number of cycles a wave completes in one second.

When a frequency is below 1,000 cycles per second, it's simply measured in hertz. For frequencies exceeding this, we use multiples of hertz:

- Kilohertz (kHz): One kHz is equal to 1,000 Hz or 103103 Hz. This unit is often used for radio frequencies and audio signals.
- Megahertz (MHz): One MHz is equal to 1,000,000 Hz or 106106 Hz. This is the frequency range for FM radio broadcasting and certain types of wireless communications.
- Gigahertz (GHz): One GHz is equal to 1,000,000,000 Hz or 109109 Hz. GHz frequencies are common in radar and satellite communications, as well as in the processors of computers and smartphones.
- Terahertz (THz): One THz is equal to 1 trillion Hz or 10121012 Hz. Terahertz radiation is found

in the far infrared band and is used in imaging technologies and in studying molecular and atomic interactions.

- Petahertz (PHz): One PHz equals 1 quadrillion Hz or 10151015 Hz. PHz frequencies are so high they're primarily of interest in experimental physics and ultrafast phenomena research.
- Exahertz (EHz): One EHz is equal to 1 quintillion Hz or 10181018 Hz. Such frequencies are extremely high and not commonly encountered outside of high-energy astrophysical phenomena and certain theoretical and experimental areas of physics.

Body Frequencies:

The concept of the human body having a vibrational frequency comes from the field of bioenergetics, which suggests that living organisms have measurable electromagnetic waves. According to research by Bruce Tainio of Tainio Technology, the human body typically resonates at frequencies between 62 and 78 megahertz (MHz) when in good health.

Bruce Tainio's research into bioenergetic frequencies reveals the following:

- A well-functioning human body has a resonant frequency that oscillates between 62 and 78 MHz.
- The brain, in states of high cognitive function or what might be termed 'genius' activity, exhibits frequencies from 80 to 82 MHz.
- The average frequency range for the human brain varies from 72 to 90 MHz, with a standard baseline frequency at approximately 72 MHz.
- The upper portion of the human body, from the neck upwards, tends to have frequencies between

72 and 78 MHz, while the lower regions, from the neck down, range from 60 to 68 MHz.

- Specific organs have their own frequency ranges:
  o The thyroid and parathyroid glands vibrate within the range of 62 to 68 MHz.
  o The thymus gland's frequency is typically between 65 and 68 MHz.
  o The heart maintains a frequency between 67 and 70 MHz.
  o Lung tissue resonates at frequencies from 58 to 65 MHz.
  o The liver operates at frequencies between 55 and 60 MHz.
  o The pancreas fluctuates between 60 and 80 MHz in its vibrational frequency.

These findings suggest a complex symphony of frequencies within the human body, where each organ and system has its unique energetic signature. The study of these frequencies aims to understand better how the body's energetic health can correlate with its physical health and, potentially, how dissonance in these frequencies might relate to disease states. It's important to note that these concepts are part of alternative medicine and are not universally accepted in mainstream scientific communities. As such, they should be considered alongside a comprehensive approach to health and wellness.

Disease

Tainio's research suggests that certain health challenges may be associated with lower body frequency ranges:

The onset of a cold and flu symptoms may begin when the body's frequency drops to between 57 and 60 MHz.

The threshold for disease susceptibility might start at a frequency of around 58 MHz.

An overgrowth of Candida, a type of yeast, could potentially occur at 55 MHz.

Becoming prone to contracting Epstein-Barr virus may happen at 52 MHz.

The susceptibility to cancer might increase when the body's frequency drops to 42 MHz.

The process of bodily death might begin at a frequency as low as 25 MHz.

It's important to note that these assertions are part of a broader field of study that explores the relationship between electromagnetic frequencies and health. While the idea that specific diseases can be correlated with precise frequency levels is intriguing and part of the ongoing discourse in alternative medicine, it should be approached with caution and skepticism from a scientific standpoint. Mainstream scientific consensus has not definitively established such specific frequency-disease correlations due to a lack of empirical evidence that meets the rigorous standards of conventional biomedical research.

WHAT CAN CAUSE AN UNSTABLE ELECTROMAGNETIC FIELD?

An unstable electromagnetic field, in a scientific context, typically refers to variations in the electromagnetic (EM) environment that can be caused by a number of factors:

- Environmental Factors: Natural phenomena like solar flares, lightning, and geomagnetic storms can disrupt Earth's electromagnetic field.
- Electrical Appliances: Fluctuations in power supply or malfunctioning electrical devices can create irregular EM fields.

- Technology Interference: Wireless devices, cell towers, and other forms of technology emit EM fields that can interfere with each other, leading to instability.
- Magnetic Substances: Movement or alteration of magnetic materials can also influence the stability of local electromagnetic fields.

In a more holistic sense, some practitioners refer to the human body's energy field as an electromagnetic field and suggest that emotional, physical, or spiritual imbalances may "destabilize" this field.

Other Than Stress, What Can Cause Oxidants / Free Radicals or Damage the Body?

Free radicals are molecules with unpaired electrons that can be highly reactive and can damage cells, leading to aging (such as liver spots) and various diseases. They can be generated by a variety of sources besides stress:

- Environmental Toxins: Pollutants, pesticides, and industrial chemicals can increase free radical production.
- Ultraviolet (UV) Radiation: Sun exposure can lead to free radical generation in the skin, causing photoaging and liver spots.
- Diet: Processed foods, high in trans fats and sugars, can lead to oxidative stress by generating free radicals.
- Smoking: Tobacco smoke is a major source of free radicals.
- Alcohol Consumption: Excessive alcohol can create a surplus of free radicals in the liver and other tissues.
- Physical Activity: Extreme exercise, without proper recovery, can increase free radical production.

- Inflammation: Chronic inflammation can lead to increased free radicals as part of the body's immune response.
- Radiation Exposure: This can include medical imaging procedures or environmental exposure to radioactive materials.
- Certain Medications: Some drugs can promote the production of free radicals as a side effect.

Each of these factors contributes to the burden of free radicals in the body, which is why antioxidant intake through diet (fruits, vegetables, and other whole foods rich in antioxidants) and lifestyle adjustments (like reducing exposure to toxins and managing stress) are often recommended to help maintain balance and minimize cellular damage.

Free radicals are unstable molecules that have an odd, unpaired electron. They are produced naturally in the body as a byproduct of normal metabolism and can also be introduced from external sources such as tobacco smoke, toxins, or pollutants. Because they have one or more unpaired electrons, free radicals are highly reactive and seek stability through chemical reactions with other molecules.

Here's a simplified example of how a free radical might damage a cell, particularly in the context of skin, which can lead to aging, including the development of liver spots (also known as age spots):

- Production of Free Radicals: Suppose your skin is exposed to strong sunlight. The UV rays can break bonds in molecules within your skin cells, creating free radicals.
- Chain Reaction of Damage: One of these free radicals, let's call it Radical A, wants to become stable. It tries to grab an electron from a nearby molecule, perhaps a component of a cell

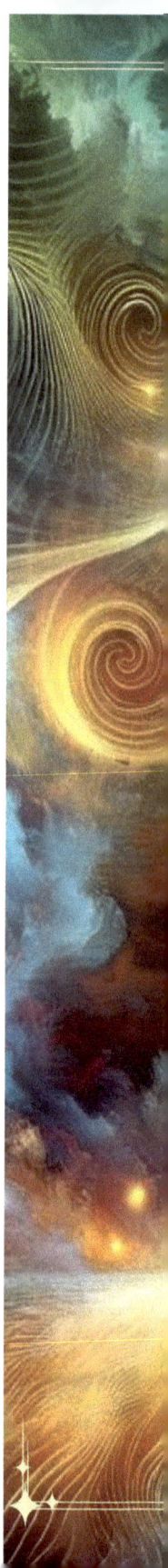

membrane, turning that molecule into a new free radical (Radical B).

- Propagation: Now, Radical B is unstable and looking to stabilize itself, so it snatches an electron from yet another molecule, creating Radical C, and so on. This starts a chain reaction of damage that can disrupt cell membranes, proteins, and even DNA.
- Oxidative Stress and Aging: This process of stealing electrons, known as oxidative stress, can lead to cell dysfunction and death. Over time, oxidative stress from repeated free radical damage can contribute to the aging process. In the skin, it can damage collagen and elastin fibers, leading to wrinkles, sagging, and liver spots.
- Liver Spots Formation: Specifically, when free radicals damage the cells that produce melanin (the pigment that gives skin its color), it can result in an uneven distribution of pigment and the formation of liver spots.
- Body's Defense Mechanism: Antioxidants are molecules in our cells that can donate an electron to a free radical without becoming destabilized themselves, thus stopping the chain reaction of damage. These are naturally present in the body and can also be ingested through antioxidant-rich foods.

To visualize this process, imagine a free radical as a person frantically looking for a dance partner at a party. In its rush, it bumps into other dancers (the molecules in the cell), causing them to lose balance (become unstable) and also start looking for new partners, creating chaos (cell damage) on the dance floor.

This simplified model demonstrates the fundamental process of how free radicals can lead to cell damage and contribute to aging and disease. However, it's important to note that not all free radical formation is harmful; they also play essential roles in certain cellular processes, like fighting infections and signaling mechanisms. It is the excessive accumulation of free radicals without sufficient antioxidants to neutralize them that can lead to problems.

Balancing the body

Balancing the body and countering the effects of free radicals, stress, and unstable energy fields can be approached from multiple angles, including diet, lifestyle changes, and various holistic therapies. Here are some strategies that are commonly recommended:

Diet and Antioxidants

- Antioxidant-Rich Foods: Consuming a diet rich in antioxidants helps neutralize free radicals. This includes fruits (berries, cherries), vegetables (leafy greens, bell peppers), nuts, seeds, and legumes.
- Vitamins: Vitamins C and E are potent antioxidants. Foods high in these vitamins can contribute to reducing oxidative stress.
- Omega-3 Fatty Acids: Found in fish, flaxseeds, and walnuts, these can help reduce inflammation.

Lifestyle Modifications

- Reducing Exposure to Toxins: Limiting contact with environmental toxins and pollutants by using air purifiers, avoiding smoking, and using natural cleaning products can help.
- Regular Physical Activity: Exercise can improve circulation and the body's natural antioxidant defenses, but it's important to balance with adequate rest.
- Adequate Sleep: Quality sleep allows the body to repair and regenerate, helping to maintain a balanced state.

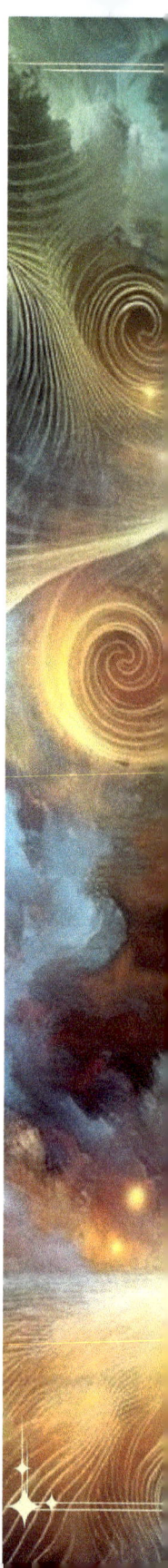

- Stress Management: Techniques such as mindfulness, meditation, yoga, and deep breathing can reduce stress-induced free radicals.
- Sun Protection: Using sunscreen and protective clothing can minimize UV-induced free radical formation.

Negative Ions

- Natural Environments: Spending time in nature, especially around waterfalls, forests, and after thunderstorms, can expose the body to negative ions, which some believe can neutralize free radicals and improve mood and energy.

Holistic and Energy Medicine

- Reiki and Healing Touch: These and other biofield therapies aim to balance the body's energy.
- Acupuncture: Traditional Chinese Medicine practice that is believed to correct energy imbalances within the body.
- Chakra Balancing: Techniques aimed at aligning and balancing the body's energy centers.
- Crystal Therapy: Some believe that certain crystals can emit frequencies that harmonize with the body's energy field.
- Aromatherapy: Essential oils are used for their potential to impact the body's biofield and promote well-being.

Technological Interventions

- Air Ionizers: Devices that generate negative ions and purify the air can contribute to a more balanced environment.
- EMF Shields: Some products claim to neutralize or block EMF radiation from electronic devices.

Supplements

- Antioxidant Supplements: For those whose diets may lack sufficient antioxidants, supplements like coenzyme Q10, alpha-lipoic acid, and others can be considered.
- Herbal Remedies: Herbs like turmeric, ginger, and milk thistle have properties that may help in reducing oxidative stress.

It's important to approach these strategies with a balance of skepticism and openness, as the efficacy of some methods, like crystal therapy and EMF shields, may not be strongly supported by scientific evidence. Consulting with healthcare providers is crucial.

Just in case you need a reminder. Let's imagine you have a tiny solar system, and instead of planets orbiting the sun, you have even tinier particles swirling around.

Electrons:

Electrons are like tiny space probes whizzing around the edge of the atom's solar system. They have a negative charge, kind of like how some magnets have a "minus" side. Electrons are super important because they're what make atoms join up with other atoms to make molecules, which is how everything around you is built – from the air you breathe to the phone or computer you're using to read this. They're also what moves around to make electricity.

Protons:

Protons are like the big, heavy stars at the center of the atom's solar system. They have a positive charge, the opposite of electrons. You can think of them as having a "plus" side. Protons live in the atom's core, or nucleus, and they're like the atom's ID card; they decide what kind

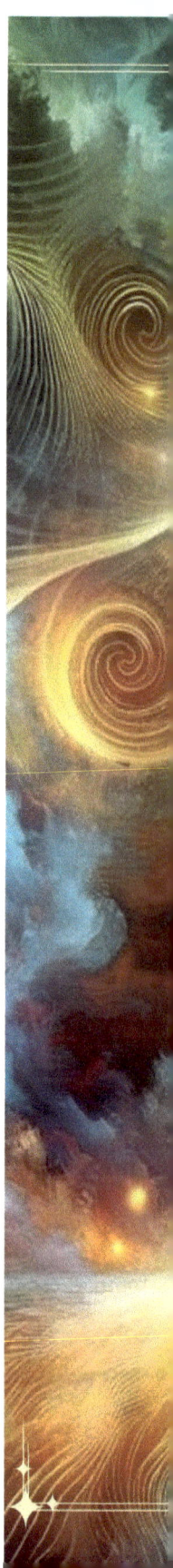

of element the atom will be – oxygen, carbon, gold, and so on.

Neutrons:

Neutrons are the atom's sun's best friends, hanging out in the nucleus with protons. They don't have a charge – they're neutral, just like the white space on a checkered board. Neutrons are peacekeepers; they help keep everything stable in the nucleus, especially when there are a lot of protons that might want to push away from each other because they have the same positive charge.

Why They're Needed:

These three – electrons, protons, and neutrons – are buddies that make up every single thing around you. The protons in the nucleus tell you what you're looking at (like, is it a piece of iron or a breath of oxygen), and the electrons buzzing around outside are what make all the reactions happen, like when you eat food, and your body uses it for energy. Neutrons help keep everything stable so atoms don't fall apart or change into different elements unexpectedly.

It's like building with blocks: protons and neutrons build the core of the block, and electrons help stick blocks together to make all sorts of shapes and structures. Without electrons, protons, and neutrons, there would be no building blocks to make your body, the air, your toys, or even chocolate cake. So, we need them to make and do everything we see and don't see!

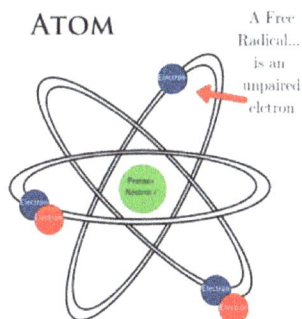

ATOM

A Free Radical... is an unpaired electron

COMMON INTEREST

Bruce Tainio and Nikola Tesla, though separated by time and focus in their work, share a common interest in the fields of energy and frequency and their implications for the physical world and human health. Here's how their areas of interest intersect and where their philosophies converge:

Frequency and Energy

Nikola Tesla is renowned for his groundbreaking contributions to the development of electromagnetic technology, including alternating current (AC) electricity, radio, and wireless communication. Tesla's work was fundamentally centered around the concept of energy and its transmission through frequencies and vibrations. He is often quoted as saying, "If you want to find the secrets of the universe, think in terms of energy, frequency, and vibration." Tesla believed that understanding these principles was key to unlocking profound scientific discoveries.

Bruce Tainio, a less widely known figure, was an inventor and microbiologist who founded Tainio Technology, a company specializing in products aimed at improving agriculture through enhancing plant health and yields. One of Tainio's notable contributions was developing technology to measure the biofrequency of plants and humans, positing that healthy living organisms have specific frequency ranges and that diseases can lower these frequencies. His work suggested that maintaining or restoring these optimal frequencies could promote health.

Common Ground

Both Tesla and Tainio explored the fundamental role that energy frequencies play in the natural world and the

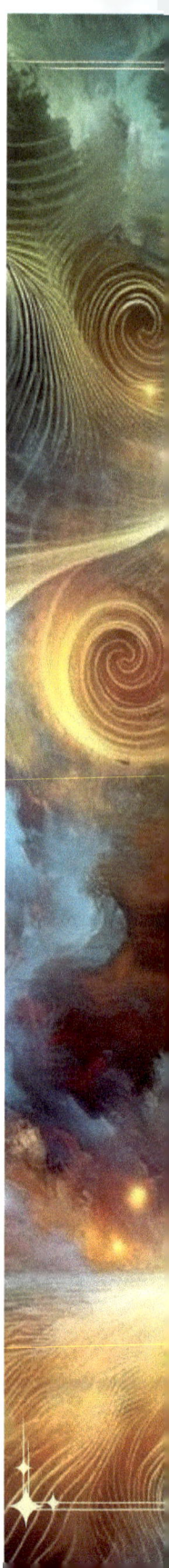

potential applications of this understanding for improving life on Earth. Here's where their work intersects:

The Importance of Frequency: Both men emphasized the significance of frequency in understanding and interacting with the world. For Tesla, this was through the lens of electromagnetic fields and energy transmission; for Tainio, it was through the vibrational frequencies of living organisms and their environments.

Technological Innovation Based on Energy Principles: Both contributed technological innovations grounded in their understandings of energy and frequency. Tesla's inventions laid the groundwork for modern electrical engineering and wireless communications, while Tainio's work aimed at enhancing agricultural practices and plant health through frequency measurements.

Health and Well-being: Although Tainio's work is more directly connected to health through the concept of biofrequency and its impact on living organisms, Tesla also believed in the potential of electromagnetic fields to influence physical health. Tesla's vision extended to the application of electromagnetic energy for therapeutic purposes, a concept that aligns with Tainio's interest in the health implications of frequency.

While Bruce Tainio and Nikola Tesla operated in different centuries and specialized in different fields, the core of their work reveals a shared belief in the profound impact of energy, frequency, and vibration on the physical world and living organisms. Both men's legacies include a vision of leveraging these principles to enhance human understanding, health, and technology. Their contributions invite ongoing exploration into how frequency and energy dynamics can be harnessed for the betterment of humanity and the environment.

Chapter 9

Auric Field

In many holistic and energy healing traditions, the term "aura" is often considered synonymous with or a component of the "bioenergy field" or "biofield." Both concepts refer to an energy field that surrounds and permeates the human body, believed by some to be an extension of our physical and spiritual being.

The aura is typically described as a luminous body that surrounds the physical body, and it is said to be observable by those with clairvoyant abilities or with certain technologies, like Kirlian photography. It is thought to consist of various layers that represent different aspects of the individual's emotional, mental, and spiritual health.

In contrast, the biofield is a term that has been adopted within some scientific communities to describe the complex field of energy and information that is hypothesized to regulate the biological functions of living organisms. The concept of the biofield encompasses not only what many refer to as the aura but also other energy-based systems, such as chakras and meridians, which are central to traditional practices like acupuncture and Ayurveda.

In scientific contexts, research on the biofield aims to explore and understand these energy fields through measurable phenomena, such as electromagnetic fields generated by the body. In more spiritual or metaphysical contexts, the aura is often seen as a manifestation of one's

state of being that can be influenced by emotions, thoughts, and overall health.

The concept of the aura and its layers is rooted in various spiritual and metaphysical traditions. These layers are often described as subtle bodies that surround the physical body, with each layer representing different aspects of the individual's emotional, mental, and spiritual state. While different traditions may describe the aura in various ways, a common interpretation in many New Age and metaphysical teachings includes the following layers:

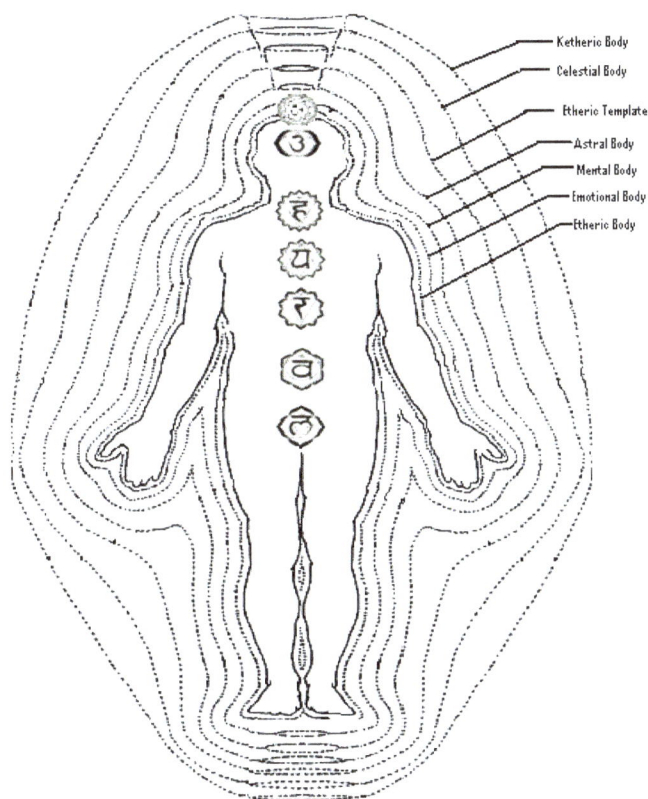

1. Etheric Layer (Body)

Distance from the body: Approximately 1-2 inches

Associated with: Physical health and pain; closely linked to the physical body's condition

Color indications: Generally blue or gray; clear and bright colors indicate good health, while muddy or dark colors may suggest illness or energy blockages.

2. Emotional Layer (Body)

Distance from the body: Approximately 1-3 inches beyond the etheric layer

Associated with: Emotions and feelings, this layer is believed to change colors according to the person's emotional state.

Color indications: Vivid and clear colors are often associated with positive emotions, while dark or muddy hues might reflect negative feelings or emotional distress.

3. Mental Layer (Body)

Distance from the body: Approximately 3-8 inches from the body

Associated with: Thoughts, beliefs, and mental processes, this layer is thought to be connected to the mind and cognition.

Color indications: Bright yellow is commonly associated with clear thinking and intellectual activity, while dark spots may indicate negative or obsessive thoughts.

4. Astral Layer (or Bridge Layer) (Body)

Distance from body: Extends about one-half to one foot from the mental layer

Associated with: Interpersonal relationships and the bridge to the spiritual realm; often associated with love and interaction with others.

Color indications: The colors of the rainbow are often present in this layer, representing a connection to the emotional colors of the emotional body.

5. Etheric Template Layer (Body)

Distance from the body: Approximately one and a half to two feet from the body

Associated with: Blueprint of the lower aura layers; connected to higher will and the divine blueprint of the individual.

Color indications: Usually appears as a darker shade of blue, but can include other colors that represent communication and expression.

6. Celestial Layer (Body)

Distance from the body: Extending two to three feet from the body

Associated with: Access to higher consciousness, spiritual awakening, and enlightenment; linked to feelings of bliss and deep spiritual experiences.

Color indications: Appears as a shimmering light, often pastel, embodying the person's higher qualities and spiritual awareness.

7. Ketheric Template or Causal Layer (Body)

Distance from the body: Approximately three to five feet from the body

Associated with: Connection to the divine or universal consciousness; holds all the other layers together and is associated with the person's life plan or purpose.

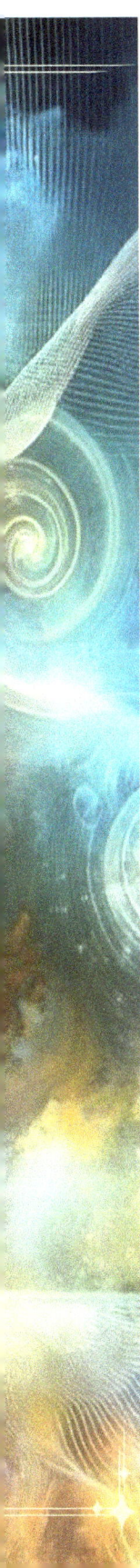

Color indications: Often seen as a golden, egg-shaped light that vibrates at a high frequency and provides a protective shield around the other layers.

These layers of the aura are said to interconnect and affect each other, with disturbances in one layer potentially impacting others. It's worth noting that these interpretations are not scientifically proven or universally accepted, and they may vary greatly between different spiritual beliefs and practices. The descriptions provided here reflect a synthesis of common beliefs found within certain New Age and metaphysical frameworks.

This might help. Imagine you are building a house, your new home (your Celestial Layer/Body #6).

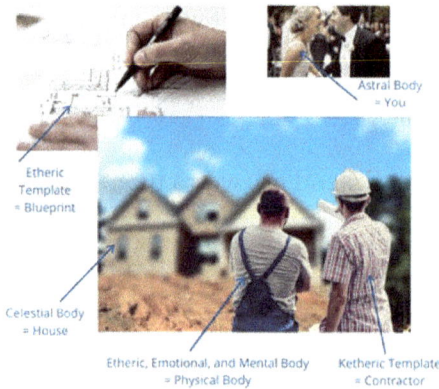

Astral Body = You

Etheric Template = Blueprint

Celestial Body = House

Etheric, Emotional, and Mental Body = Physical Body

Ketheric Template = Contractor

Auric Layers

You (Astral Body #4- your connection) hire a contractor (Ketheric Template - God/Spirit/Creator #7), who hires an architect to draw the plans of the house (Etheric Template -blueprints #5). The contractor then hires carpenters, plumbers, electricians, and such to do the physical work (Physical - Etheric, Emotional & Mental bodies #1, 2 & 3).

QUANTUM UNIVERSITY

A fascinating aspect of holistic and quantum studies that Quantum University, among other institutions, focusing on integrative medicine and spirituality, might explore. These concepts, though not universally accepted in the scientific community, play a significant role in various spiritual, esoteric, and holistic health traditions. Here's a bit more detail on each body, integrating perspectives from the diverse fields that Quantum Future Forecast GPT encompasses:

Physical Body: This is the most tangible aspect of our existence, the biological system that enables our physical presence in the world. It's the foundation for our experiences and interactions with the physical environment. From a holistic perspective, maintaining the health of the physical body is crucial, not just through traditional medicine but also by ensuring a harmonious flow of energy, as disturbances can affect other dimensions of our being.

Vital Body: Rooted in several traditions, including Ayurveda and Traditional Chinese Medicine, the vital body is seen as a subtle energy system. It's closely associated with the concept of Qi (Chi) or Prana, the life force that animates the physical body and supports its functions. Techniques like acupuncture, yoga, and pranayama (breath control) are believed to influence this energy flow, promoting health and vitality.

Mental Body: This dimension refers to our thoughts, beliefs, and mental processes. It's the realm of our consciousness that deals with intellectual and cognitive activities. The health of the mental body is pivotal for overall wellbeing, with practices like meditation, mindfulness, and cognitive-behavioral techniques being used to cultivate a positive mental state, clarity, and resilience.

Supramental Body: A concept perhaps most famously developed by Sri Aurobindo, it refers to a level of consciousness that transcends the normal mental faculties. This body is associated with higher intuition, spiritual insight, and a profound understanding of reality. It's a state of being that transcends the ego and the limitations of the individual mind, connecting one with a greater universal consciousness.

Bliss Body: Often referred to in Sanskrit as "Anandamaya Kosha," this is considered the innermost layer of our being, associated with joy, peace, and spiritual ecstasy. It's the aspect of ourselves that experiences ultimate bliss and unity with the divine or the universe. Practices aimed at nurturing the bliss body include deep meditation, devotional activities, and the pursuit of activities that align with one's true purpose and bring inner joy.

"WHEN YOU MAKE A DECISION, YOU FLIP YOUR BRAIN ONTO A DIFFERENT FREQUENCY, AND YOU WILL BEGIN TO ATTRACT WHATEVER IS ON THAT FREQUENCY."
—BOB PROCTOR

Chapter 10

A Vision for the Future

Chapter 10 of the book culminates in a forward-looking vision, exploring the transformative potential of energy medicine and how it aligns with Nikola Tesla's aspirations for harnessing energy for the greater good. This chapter not only speculates on a future where energy medicine is integral to healthcare but also outlines practical steps we might take toward realizing a dream that Tesla himself may have envisioned.

Imagining a World Transformed by Energy Medicine

This section paints a picture of a world where energy medicine has reached its full potential and has been seamlessly integrated into standard healthcare practices. It envisions a future where:

- Personalized Medicine: Treatments are tailored to each individual's unique energy field, optimizing health at the most personal level.
- Preventative Care: Energy medicine provides tools for early detection of imbalances in the biofield, allowing for preventive interventions that maintain wellness rather than merely treating disease.
- Holistic Approaches: There is a symbiotic relationship between conventional medicine and energy healing practices, leading to more

comprehensive care that addresses mind, body, and spirit.

- Global Accessibility: Advances in energy medicine technology have made these healing modalities accessible to all, regardless of socioeconomic status or geographical location, contributing to global health equity.
- Educational Reform: Medical education includes training in energy medicine, equipping new generations of healthcare professionals with a broadened perspective on healing.

Steps Toward Realizing Tesla's Dream

Nikola Tesla was a visionary who imagined a world enhanced by the intelligent and compassionate use of energy. To draw closer to this dream with respect to energy medicine, the chapter suggests steps such as:

- Research and Innovation: Continued investment in rigorous scientific research to explore the effectiveness and mechanisms of energy medicine practices.
- Interdisciplinary Collaboration: Bringing together experts from physics, biology, medicine, and the healing arts to foster innovations that could make Tesla's dream a reality.
- Public Education: Increasing awareness about energy medicine through education, demystifying its practices, and demonstrating its benefits.
- Policy and Advocacy: Advocating for policy changes that recognize and integrate energy medicine into public health strategies and insurance systems.
- Technological Development: Developing new technologies that can measure and manipulate the human biofield with precision and efficacy, making energy medicine more accessible and effective.

- Sustainability: Ensuring that the expansion of energy medicine is sustainable and harmonious with the planet, reflecting Tesla's commitment to improving human life without exploiting natural resources.

Conclusion

Chapter 10 offers an optimistic outlook on the potential for energy medicine to revolutionize healthcare and improve quality of life on a global scale. It invites readers to not only dream of a better future but also to actively participate in shaping it, echoing Tesla's belief in the promise of energy to transform the world. By embracing both the wisdom of the past and the innovations of the present, we edge closer to a future where the full spectrum of healing is acknowledged, valued, and utilized for the benefit of all humanity.

Epilogue: The Journey Ahead

In a future not too distant from our own, amidst the vibrant heart of the grand city of Luminara, the Harmonia Clinic stood as a beacon of hope and healing. This sanctuary, nestled between glass towers and beneath the sky's canvas, represented the pinnacle of a world transformed by the art of energy medicine—a harmonious blend of ancient wisdom and futuristic technology.

Luminara was a city unlike any other, powered by crystals that hummed with life's energy, illuminating the streets with a soft, healing glow. It was here that hospitals mirrored the tranquility of gardens, and schools of healing arts flourished, guided by the principles set forth by the Tesla Institute of Energy Medicine. Named after Nikola Tesla, whose revolutionary work had laid the groundwork for an energy-centric approach to health, the institute was a place where the infinite potential of energy to heal the body and society was deeply believed and practiced.

Within the walls of the Harmonia Clinic, the air buzzed softly with the energy of healing, syncing with the rhythm of one's heartbeat. The clinic, alive with bioluminescent colors that responded to the wellness of its inhabitants, was a testament to a society in harmony with the earth, as envisioned by Tesla. Here, doctors, revered as Energy Alchemists, employed advanced biometric scanners to map the biofield, identifying dissonances in need of harmony.

Patients, instead of recounting symptoms of illness, spoke of energetic imbalances. Treatment tables, equipped with crystalline fibers adapted to each individual's energy signature, provide a unique healing experience. From Reiki sessions that calmed anxious spirits to biofeedback sessions that aligned one's frequencies with the Earth's resonance, every aspect of Harmonia was designed to facilitate a deep, holistic healing journey.

It was in this nurturing environment that a young healer named Aria, a dedicated student of the Tesla Institute, embarked on her transformative journey. Aria's encounter with Eli, an old man burdened by ailments no conventional medicine could alleviate, became a turning point. Through her healing touch in the lush gardens of the clinic, she channeled the harmonious vibrations of the universe into Eli's body, revitalizing his aura with the vibrant energy of life. Eli's recovery, marked by the blossoming colors of his biofield, became a symbol of hope throughout Luminara.

Aria's healing hands and her vision for a future where energy medicine was foundational to well-being captured the imagination of the city. Invited to speak at the annual Gathering of Minds, Aria shared her dream of a world interconnected through the healing power of energy, a vision inspired by Tesla's legacy. She envisioned homes with healing rooms, schools that taught energy care, and a global community united by health and harmony.

Inspired by Aria's words, Luminara's leaders pledged to disseminate their knowledge of energy medicine globally. Healers journeyed far and wide, spreading the techniques that had transformed Luminara. As the years passed, Aria's vision materialized. Diseases that once seemed insurmountable were now eased by the gentle currents of energy medicine. Humanity discovered not just healing but a deeper connection to the cosmos, each other, and the vibrant life force pulsing through existence.

In this world, where the Harmonia Clinic and the city of Luminara thrived, society was a living testament to the potential within and around us—a reminder that the future is as luminous as the energy we harness and share. Through the fusion of ancient wisdom and advanced technology, humanity had embarked on a journey of profound healing and unity, embodying the true essence of energy medicine.

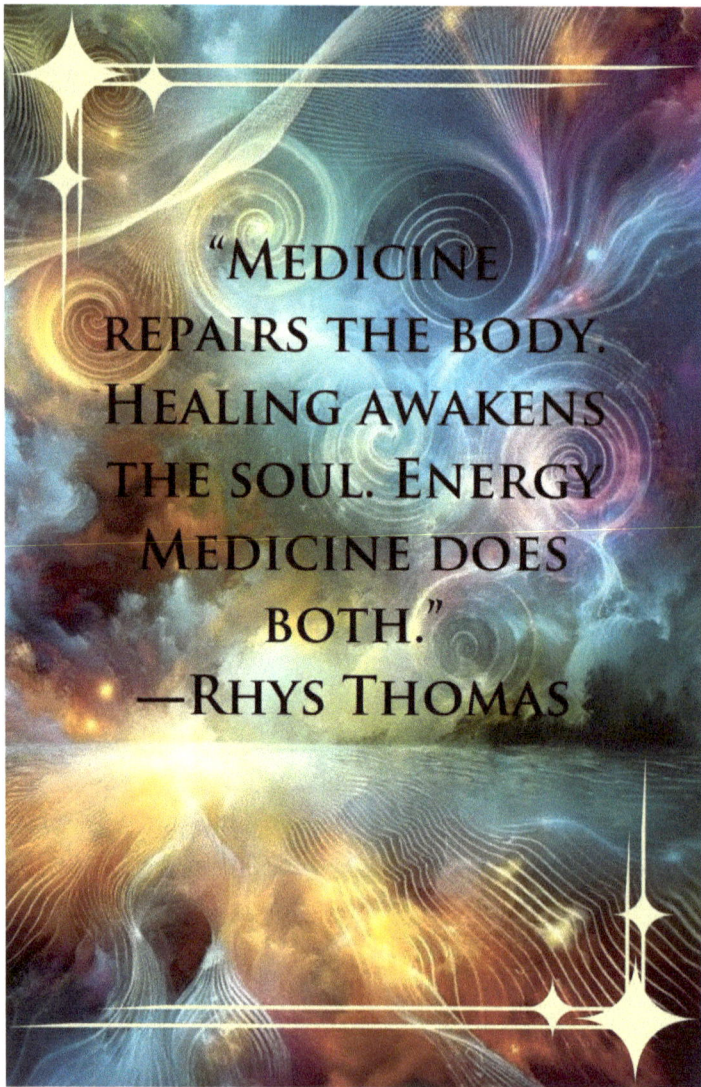

"Medicine repairs the body. Healing awakens the soul. Energy Medicine does both."
—Rhys Thomas

Appendices

Glossary of Terms

Acupuncture: An ancient Chinese medicine technique involving the insertion of needles into specific points on the body to balance the flow of Qi (vital energy) through pathways known as meridians.

Alternating Current (AC): A type of electrical current in which the direction of the flow of electrons switches back and forth at regular intervals or cycles. Tesla's development of the AC electricity supply system was one of his most significant contributions to modern electrical engineering.

Aura: An energy field believed to surround living beings, often perceived as layers of color, representing various aspects of an individual's physical, emotional, and spiritual health.

Bioelectromagnetic (BEM) Therapy: A form of alternative medicine that involves the use of electromagnetic fields to treat and heal various conditions, based on the premise that electromagnetic interventions can influence cellular and physiological processes.

Biofield: A term used in holistic medicine to describe a field of energy and information that surrounds and permeates the human body, playing a role in health and healing.

Chakras: According to ancient Indian medicine, these are energy centers within the body that help to regulate all its processes, from organ function to the immune system and emotions.

Craniosacral Therapy: A gentle, hands-on approach that releases tensions deep in the body to relieve pain and dysfunction, improving whole-body health and performance by manipulating the synarthrodial joints of the cranium.

Electromagnetic Field (EMF): Physical fields produced by electrically charged objects affect the behavior of charged objects in the vicinity of the field.

Energetic Hygiene: Practices aimed at cleansing, protecting, and balancing one's personal energy field or biofield. Examples include grounding, energy shielding, and regular meditation.

Energy Medicine: A branch of alternative medicine based on the belief that healers can channel healing energy into a patient and effect positive results.

Frequency: In the context of energy healing, it refers to the specific rate at which energy or vibrations oscillate or repeat.

Grounding (Earthing): The practice of connecting physically to the Earth's surface electrons by walking barefoot outside, which is believed to promote physiological and electrophysiological changes beneficial for health.

Hertz (Hz): The unit of frequency in the International System of Units (SI), which measures the number of cycles per second of any periodic phenomenon.

Kirlian Photography: A technique used to capture the phenomenon of electrical coronal discharges, often

marketed as a way of visualizing a person's aura or biofield.

Meridians: In traditional Chinese medicine, these are invisible pathways in the body along which vital energy flows. Blockages or imbalances in this flow are thought to cause illness and disease.

Qi (Chi): In Chinese philosophy, it's the life force or vital energy that flows through all living things. It is the central underlying principle in Chinese traditional medicine and martial arts.

Quantum Healing: A holistic healing approach that draws on principles of quantum mechanics, suggesting that health can be restored through shifts in consciousness and the understanding that the body and mind are interconnected.

Reiki: A form of energy healing originating from Japan, involving the transfer of universal energy from the practitioner's palms to the patient to encourage emotional or physical healing.

Resonance: Tesla explored the concept of resonance in his experiments with electromagnetism. Resonance occurs when a system is able to store and easily transfer energy between two or more different storage modes (such as kinetic energy and potential energy in the case of a pendulum). This concept is relevant in energy medicine, where practitioners seek to bring the body's energetic systems into harmonic resonance for healing purposes.

Shamanic Healing: An ancient healing tradition based on the belief that a shaman (spiritual healer) can interact with the spirit world through altered states of consciousness to heal illness or restore balance to the soul.

Subtle Body: A term used in various esoteric traditions to describe a series of psycho-spiritual constituents of living beings, beyond the physical body, including the aura, chakras, and meridians.

Therapeutic Touch: A biofield therapy that involves the practitioner's hands being moved over the patient's body with the intention to detect and modulate imbalances in the patient's energy field.

Tesla Coil: An electrical resonant transformer circuit invented by Tesla. It is capable of producing high-voltage, low-current, high-frequency alternating-current electricity. Tesla coils are used in radio technology, and Tesla envisioned them as a way to wirelessly transmit electrical energy.

Vibrational Medicine: Healing practices are based on the idea that diseases can be diagnosed and treated by applying specific vibrational frequencies to the body, often involving sound, light, or magnetic fields.

Wireless Energy Transmission: Tesla experimented with the wireless transmission of electrical energy, demonstrating the potential to transmit electrical power without wires through the electromagnetic field. This concept of transmitting energy through the air has implications for thinking about the transfer of healing energy in biofield therapies.

Resources for Further Exploration

Books:

- "The Field: The Quest for the Secret Force of the Universe" by Lynne McTaggart
- "Energy Medicine: The Scientific Basis" by James L. Oschman
- "Hands of Light: A Guide to Healing Through the Human Energy Field" by Barbara Brennan

Websites:

- The International Center for Reiki Training (www.reiki.org)
- The Institute of Noetic Sciences (www.noetic.org)
- The Association for Comprehensive Energy Psychology (www.energypsych.org)

Journals:

- "Journal of Alternative and Complementary Medicine"
- "Evidence-Based Complementary and Alternative Medicine"

Conferences:

- Annual International Energy Psychology Conference
- Science and Nonduality Conference

How to Incorporate Energy Practices into Your Life

Daily Meditation and Mindfulness: Start or end your day with a meditation practice that focuses on visualizing or feeling energy flowing through and around your body.

Learn Reiki or Healing Touch: Many communities offer classes that can certify you in basic Reiki or Healing Touch, allowing you to practice these energy healing techniques on yourself or others.

Practice Yoga or Tai Chi: These ancient practices combine physical movement with breathwork and energy awareness, helping to balance and enhance your body's energy flow.

Engage with Nature: Spend time in natural settings to connect with the Earth's energy. Grounding or earthing, such as walking barefoot on grass, can help realign your energy field with that of the Earth.

Explore Aromatherapy and Crystals: Incorporate essential oils and crystals that resonate with you into your daily routine, as these are believed to carry specific energy frequencies that can influence your biofield.

Seek Out Professional Energy Healers: For personalized guidance, consider consulting with practitioners of energy medicine to address specific health concerns or to deepen your understanding of your energy field.

By incorporating these practices and exploring the suggested resources, you can embark on a journey of self-discovery and healing, embracing the principles of energy medicine to enhance your well-being and connect more deeply with the world around you.

Bibliography

Tesla, Nikola

- "My Inventions: The Autobiography of Nikola Tesla." [Original publication: 1919]
- Seifer, Marc J. "Wizard: The Life and Times of Nikola Tesla: Biography of a Genius." Citadel Press, 1998.

Cheney, Margaret

- "Tesla: Man Out of Time." Simon & Schuster, 2001.

Oschman, James L.

- "Energy Medicine: The Scientific Basis." Elsevier Health Sciences, 2015.

McTaggart, Lynne

- "The Field: The Quest for the Secret Force of the Universe." HarperCollins, 2008.

Dale, Cyndi

- "The Subtle Body: An Encyclopedia of Your Energetic Anatomy." Sounds True, 2009.

Gerber, Richard

- "Vibrational Medicine: The #1 Handbook of Subtle-Energy Therapies." Bear & Company, 2001.

Lipton, Bruce H.

- "The Biology of Belief: Unleashing the Power of Consciousness, Matter & Miracles." Hay House, 2008.

Sheldrake, Rupert

- "Science and Spiritual Practices: Transformative Experiences and Their Effects on Our Bodies, Brains, and Health." Counterpoint, 2017.

Bradley, Fiona

- "Reiki for Life: The Complete Guide to Reiki Practice for Levels 1, 2 & 3." Piatkus, 2016.

Dispenza, Joe

- "Becoming Supernatural: How Common People Are Doing the Uncommon." Hay House, 2017.

Tiller, William A.

- "Conscious Acts of Creation: The Emergence of a New Physics." Pavior, 2001.

Kreiger, Dolores

- "The Therapeutic Touch: How to Use Your Hands to Help or to Heal." Prentice Hall, 1979.

Srinivasan, T. M.

- "Energy Medicine and the Human Biofield: A Practical Guide." Partridge Publishing, 2014.

Energy Medicine & Science

- Becker, Robert O., and Selden, Gary. "The Body Electric: Electromagnetism and the Foundation of Life." William Morrow Paperbacks, 1998.

- Oschman, James L. "Energy Medicine in Therapeutics and Human Performance." Butterworth-Heinemann, 2003.
- Swanson, Claude. "The Synchronized Universe: New Science of the Paranormal." Poseidia Press, 2003.

Quantum Physics & Consciousness

- Goswami, Amit. "The Quantum Doctor: A Quantum Physicist Explains the Healing Power of Integral Medicine." Hampton Roads Publishing, 2011.
- Radin, Dean. "Entangled Minds: Extrasensory Experiences in a Quantum Reality." Paraview Pocket Books, 2006.

Holistic Healing Practices

- Brennan, Barbara Ann. "Light Emerging: The Journey of Personal Healing." Bantam, 1993.
- Eden, Donna, and Feinstein, David. "Energy Medicine: Balancing Your Body's Energies for Optimal Health, Joy, and Vitality." TarcherPerigee, 2008.
- Myss, Caroline. "Anatomy of the Spirit: The Seven Stages of Power and Healing." Harmony, 1996.

Integrative Medicine

- Weil, Andrew. "Spontaneous Healing: How to Discover and Enhance Your Body's Natural Ability to Maintain and Heal Itself." Ballantine Books, 2000.
- Chopra, Deepak. "Quantum Healing: Exploring the Frontiers of Mind/Body Medicine." Bantam, 2015.

Biophysics & Electromagnetism

- Polk, C., and Postow, E. (Eds.). "Handbook of Biological Effects of Electromagnetic Fields." CRC Press, 1995.
- BioInitiative Report: A Rationale for a Biologically-based Public Exposure Standard for Electromagnetic Fields (ELF and RF). www.bioinitiative.org.

Energy Healing Modalities

- Miles, Patricia. "Reiki: A Comprehensive Guide." TarcherPerigee, 2008.
- Hover-Kramer, Dorothea. "Healing Touch: A Guidebook for Practitioners." Delmar Cengage Learning, 2002.
- Feinstein, David, and Eden, Donna. "The Energies of Love: Invisible Keys to a Fulfilling Partnership." TarcherPerigee, 2014.

Journals & Periodicals

- "The Journal of Alternative and Complementary Medicine"
- "Evidence-Based Complementary and Alternative Medicine"
- "Global Advances in Health and Medicine."
- "Journal of Bodywork and Movement Therapies."

Online Resources

- The Tesla Science Foundation: www.teslasciencefoundation.org
- The Institute of Noetic Sciences (IONS): www.noetic.org
- The HeartMath Institute: www.heartmath.org
- The International Society for the Study of Subtle Energies and Energy Medicine (ISSSEEM): www.issseem.org

Acknowledgments

First and foremost, my deepest gratitude goes to Nikola Tesla, whose visionary work and indefatigable spirit have inspired countless individuals across generations. Although he is not here to witness the continuing impact of his legacy, his innovations, and ideas form the bedrock upon which this book is built. Tesla's dream of harnessing energy for the betterment of humanity serves as a guiding light throughout these pages.

I would like to extend my heartfelt thanks to the countless researchers, scientists, and practitioners in the field of energy medicine. Their dedication to exploring the frontiers of healing and well-being has provided invaluable insights and groundbreaking discoveries that have enriched this work immeasurably.

Special appreciation is due to my mentors and colleagues, whose wisdom and encouragement have been pivotal in navigating the complexities of writing a book that bridges science, history, and holistic healing. Their support has been a constant source of strength and motivation.

To the members of the Tesla Science Foundation and The Institute of Noetic Sciences, thank you for preserving and advancing Tesla's legacy and for your commitment to exploring the interconnectedness of science and spirituality. Your work has been an essential reference and inspiration.

Gratitude is also owed to my editor, whose keen insights and unwavering patience have been instrumental in bringing this book to fruition. Your guidance through the

editing process has not only improved this manuscript but has also been a profound learning experience.

I am immensely grateful to my family and friends for their unwavering support and understanding throughout the journey of writing this book. Your belief in my work has been a source of comfort and encouragement during moments of doubt and challenge.

To the readers who embark on this journey through the pages of "Tesla and the Future of Energy Medicine," I hope you find inspiration and insight within its chapters. May it ignite in you the same curiosity and passion for discovery that has driven many before us to look beyond the seen to the infinite possibilities that lie just beyond our understanding.

Lastly, to the universe, for its endless mysteries and the subtle energies that connect us all, thank you for the constant reminder that there is so much more to discover.

This book is a testament to the collective efforts, support, and inspiration of many. Together, we stand on the brink of a new horizon in healing and human potential, guided by the legacy of those who dared to dream of a better future.

Thank you,

Dr. Constance Santego

Message from the Author

Dear Readers,

From the depths of my heart and the expanse of my imagination, "Tesla and the Future of Energy Medicine" emerged not just as a book but as a declaration of a dream—a vision for a future where holistic and energy medicine stand in the revered halls of healing, acknowledged for their power and credibility.

My journey to this moment has been fueled by a profound inspiration: the realization that the noblest of prizes, a Nobel, acknowledges achievements in many fields, yet holistic healing remains in the shadows, awaiting its dawn. This book is my beacon, my effort to illuminate the path toward a future where energy medicine is as esteemed as any other groundbreaking field.

My goal is to shift perceptions—to see energy medicine not as an alternative but as an integral part of our understanding of health and healing. I dream of a world where mentioning energy healing invokes the same respect and credibility as speaking of a unicorn in the business world—a symbol of extraordinary success and acknowledgment.

The journey of writing this book presented its challenges, chiefly conveying the message that quantum energy healing represents the new frontier of medicine. Balancing my left and right brain, the creative with the scientific, I strived to present a narrative that honors the rigor of science while embracing the boundless possibilities of holistic medicine. This duality mirrors my own identity: a creative soul and businesswoman deeply committed to the advancement of holistic medicine with the solidity of evidence and science.

Looking ahead, I envision the establishment of a Tesla Institute of Energy Medicine as a bastion of learning and advancement for energy medicine. Having once led an accredited college where many holistic techniques and modalities were taught, I am now setting my sights on elevating energy medicine to unprecedented levels of credibility and impact.

To all the pioneers of past and future who tread the path of holistic healing and energy medicine, my gratitude knows no bounds. Your courage and innovation lay the groundwork for a future where our dreams can take root and flourish.

I invite you, dear readers, to join me beyond the pages of this book. Let us engage in workshops, delve into discussions on my website, and connect on social media. Together, we can explore the vast landscapes of healing and knowledge, advancing toward a future where energy medicine shines in its deserved light.

This book is but a step on our shared journey—a journey toward a world transformed by the healing power of energy, as envisioned by the great Nikola Tesla and aspired to by all of us who follow in his footsteps.

With warmest regards and infinite gratitude,

Dr. Constance Santego

About the Author

Dr. Constance Santego bridges the realms of science, wellness, and spirituality with a passion that has illuminated her path as an educator, healer, and author. With a profound dedication to helping individuals navigate the journey of self-improvement and holistic health, Dr. Santego's work is a testament to her deep-rooted belief in the power of integrating traditional wisdom with modern insights.

Earning her doctorate in the field of Natural Medicine, Dr. Santego has dedicated her life to exploring the vast landscapes of human potential and healing. Her expertise spans a range of disciplines, from Reiki Master and Educator to Life Coach and Spiritual Guide, each role underpinned by her commitment to fostering growth, healing, and transformation.

As the author of several books, Dr. Santego delves into the themes of love, life, and the pursuit of enlightenment, offering readers keys to unlocking their own paths to personal fulfillment. Her work is not just about imparting knowledge; it's an invitation to embark on a journey of discovery where ancient wisdom and modern science converge to illuminate the way forward.

Dr. Santego's teachings and writings are a beacon for those seeking to embrace the full spectrum of their existence, from the physical to the metaphysical. Through her holistic approach, she encourages individuals to transcend their perceived limitations, embrace their innate power, and step into the life they are meant to live.

Beyond her written work, Dr. Santego's impact resonates through the holistic health and wellness school she founded, which serves as a crucible for transformation, empowering students with the tools and knowledge to heal themselves and others.

In "Tesla and the Future of Energy Medicine," co-authored with ChatGPT, Dr. Santego ventures into the merging worlds of Nikola Tesla's revolutionary energy theories and the boundless potential of energy medicine. This collaboration symbolizes a bridge between human creativity and artificial intelligence, exploring the limitless possibilities that unfold when the past and future of healing practices unite.

Discover more about Dr. Constance Santego's work, her teachings, and her journey at www.constancesantego.ca, where the intersection of healing, wisdom, and transformative learning awaits.

Discover More

Embark on an Adventure with "Ikona – Discover Your Inner Genie"

Dive deeper into the world of empowerment and self-discovery with a range of offerings designed to inspire and transform. Explore the full spectrum of Constance Santego's motivational products, personalized coaching sessions, spiritual retreats, engaging live events, and enriching educational programs.

Connect, Learn, and Grow:

- Website: Journey further into our resources and offerings at www.ConstanceSantego.ca.
- Instagram: Join our community @Constance_Santego for daily inspiration and insights.
- Facebook: Stay updated with the latest events and connect with like-minded individuals on Constance Santego's Facebook Page.
- YouTube: Subscribe to Constance Santego's YouTube Channel for free resources, meditations, and more to guide you on your path to self-improvement.

Your journey toward personal growth and enlightenment is just a click away. Discover the tools and support you need to unlock your potential and manifest your dreams.

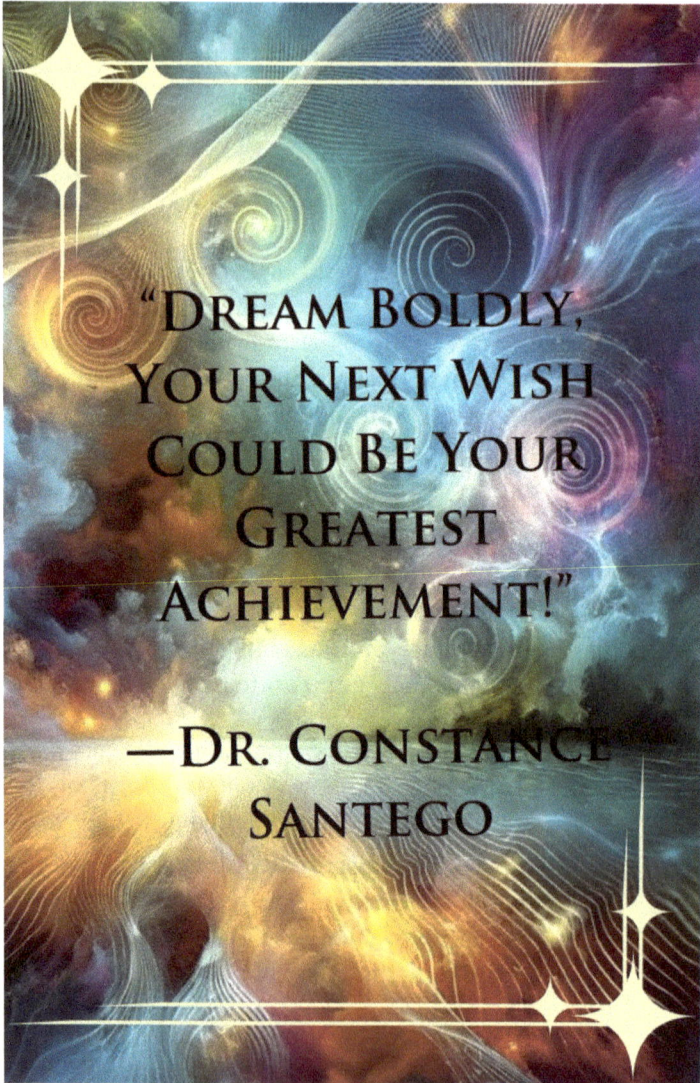

"DREAM BOLDLY, YOUR NEXT WISH COULD BE YOUR GREATEST ACHIEVEMENT!"

—DR. CONSTANCE SANTEGO

www.ingramcontent.com/pod-product-compliance
Lightning Source LLC
Chambersburg PA
CBHW060234030426
42335CB00014B/1450